A Guide for Using

The Magic School Bus® Inside a Hurricane

in the Classroom

Based on the book written by Joanna Cole

This guide written by ***Greg Young, M.S. Ed.***

Teacher Created Materials, Inc.
6421 Industry Way
Westminster, CA 92683
www.teachercreated.com

Reprinted, 2001
ISBN-1-57690-089-4
Made in U.S.A.

Edited by
Mary Kaye Taggart

Illustrated by
Howard Chaney

Cover Art by
Wendy Chang & Diane Birnbaum

Table of Contents

Introduction

The use of literature can enhance the study of science. The key to selecting these books is to check them for scientific accuracy and appropriateness for the level of the students. *The Magic School Bus*® series, written by Joanna Cole, is an outstanding example of books which can help students enjoy and learn about science. These books are delightfully written and scientifically accurate, thanks to the thorough research done by the author as she writes each of her books.

This Science/Literature Unit is directly related to *The Magic School Bus® Inside a Hurricane.* It is designed to help you present exciting lessons for your students so that they can develop their understanding and appreciation of the phenomena of hurricanes. The activities in this unit are particularly appropriate for intermediate and middle grades.

Internet Extenders

Today, many classrooms are connected to the Internet, which puts students in touch with worldwide resources. As with selecting the literature, the author of this unit has searched the Internet to find quality Web sites that directly relate to the topics covered in this book. This supplemental information helps to expand the students' knowledge of the topic, as well as make them aware of the many valuable resources to be found on the Internet. Some Web sites lend themselves to group research; others are best viewed by the entire class. If available, use a large-screen monitor when the entire class is viewing a Web site and discussing its content.

Internet Extenders are included throughout the unit. These are special sections which note outstanding Web sites and include suggestions for incorporating Web site information into the lessons. Where appropriate, a Technology Extender, such as the use of video cameras, may also be suggested in the lessons.

Although these Web sites have been carefully selected, they may not exist forever. One method of assuring that Web site information will continue to be available for a class is to use a program (if available) or browser which enables one to download and save a single Web page or entire site. Teacher Created Materials attempts to offset the ongoing problem of sites which move, "go dark," or otherwise leave the Internet after a book has been printed. If you attempt to contact a Web site listed in this unit and find that it no longer exists, check the TCM home page at www.teachercreated.com for updated URL's for this book.

About the Author Joanna Cole

Joanna Cole was born on August 11, 1944, in New York. She attended the University of Massachusetts and Indiana University before receiving her B.A. from the City College of University of New York in 1967.

Joanna Cole loved science as a child. "I always enjoyed explaining things and writing reports for school. I had a teacher who was a little like Ms. Frizzle. She loved her subject. Every week she had a child do an experiment in front of the room and I wanted to be that child every week," she recalls. It's no surprise that when she was a child Cole's favorite book was *Bugs, Insects, and Such.*

Ms. Cole has worked as an elementary school teacher, a librarian, and a children's book editor. Combining her knowledge of children's literature with her love of science, she decided to write children's books. Her first book was *Cockroaches* (1971), which she wrote because there had never been a book written about the insect before. "I had ample time to study the creature in my low-budget New York apartment!"

Teachers and children have praised Ms. Cole's ability to make science interesting and understandable. Her *Magic School Bus*® series has now made science funny as well. Cole says that before she wrote this series, she had a goal to write good science books telling stories that would be so much fun to read that readers would read them even without the science component.

Readers across the country love the *Magic School Bus*® series and enjoy following the adventures of the wacky science teacher, Ms. Frizzle. Joanna Cole works closely with Bruce Degen, the illustrator for this series, to create fascinating and scientifically accurate books for children.

At times, even a successful writer finds it scary to begin writing a new book. That was the way Ms. Cole felt before beginning to write the *Magic School Bus*® series. She says, " I couldn't work at all. I cleaned out closets, answered letters, and went shopping—anything but sit down and write. But eventually I did it, even though I was scared."

Joanna Cole says kids often write their own *Magic School Bus*® adventures. She suggests they just pick a topic and a place for a field trip. Do a lot of research about the topic. Think of a story line and make it funny. Some kids even like to put their own teachers into their stories.

The Magic School Bus® Inside a Hurricane

by Joanna Cole

(Scholastic, 1986)

(Canada, Scholastic; U.K., Scholastic Limited; AUS, Ashton Scholastic Party Limited)

Ms. Frizzle and her class are preparing for another field adventure. Most classes take field trips, but with the Friz, it is always an adventure. This time, the class is going to learn all about the weather. They are preparing for their trip by measuring and recording different aspects of the weather, such as temperature, wind speed, and rainfall. Several students are even doing reports on the air and atmosphere.

One day, before the class has even finished their experiments about air, the Friz decides it is time to go on their adventure. With parent permission slips in hand, the students board the bus and are on their way to learn about the weather. But this is no ordinary bus; this is a magic bus which can transform into a hot air balloon so that the students can get up close to the weather.

Flying high over the tree tops and mountains, the class enters clouds and learns about rain formation. Their portable weather radio warns them of a hurricane developing in the tropics, and the Friz decides to take the class right into the storm (after all, she does have parent permit slips). As the storm builds, the students learn what it takes to create a hurricane and are swept up in the process.

It is not long before trouble begins, and the magic bus/balloon begins to fall through the eye of the hurricane. Tumbling through the air, the class manages to scramble back into the bus which now transforms itself into a weather plane. Poor Arnold is left behind! He missed the bus and is now falling towards the ocean. No matter, Ms. Frizzle (truly a professional educator) marks him absent and moves ahead with the lesson.

The class and Ms. Frizzle track Arnold's progress as he is picked up by some fishermen and is escorted safely to land. The Friz steers the magic bus over to where Arnold is standing and rescues him. But the worst is not over yet. The magic bus is swept up into a tornado, and the students think they are goners!

Fortunately, Ms. Frizzle is an excellent driver, and the tornado's force dies down quickly to allow the class to return safely back to land. The bus pulls into a magic gas station before moving on to the weather station where the students meet with meteorologists and tell their tale.

Another exciting field adventure brought to the class by the wild Ms. Frizzle and her magic bus has come to an end.

Hot and Cold

Ms. Frizzle's class is preparing for their lesson on weather. The students are busy recording daily temperatures and learning how the sun influences the earth's weather.

Just what is temperature anyway? Temperature is actually a measurement of how fast the particles in air or water are moving. If these particles, also known as molecules, are moving quickly, we feel a warm or hot temperature. If these particles are moving slowly, we feel a cool or cold temperature.

Try this activity to see the relationship between temperature and the speed of particles.

Materials:

- hot water
- cold water
- blue food dye
- red food dye
- two clear drinking glasses

Procedure:

1. Fill one drinking glass with hot water and the other with cold water.
2. Allow the water to settle until it is standing still.
3. Simultaneously, drop one drop of blue food dye in the cold water and one drop of red food dye in the hot water.

Hot

Cold

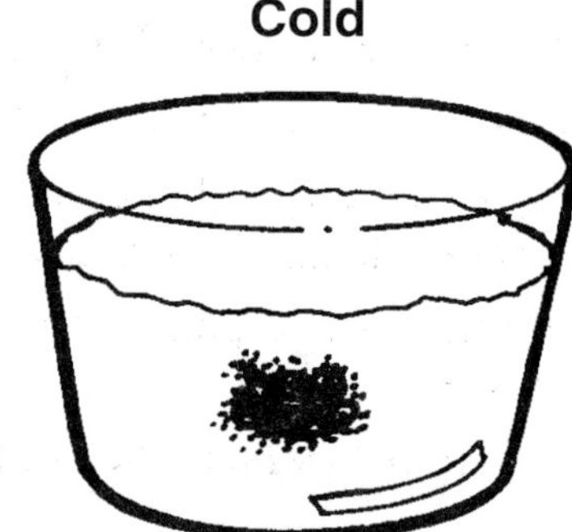

Results:

The red food dye will spread out quickly as the hot water particles/molecules move quickly. The blue dye will spread out slowly as the cold water particles/molecules move slowly.

Closure:

1. How many seconds does it take the hot water to become completely red? ____________________
2. How many seconds (minutes) does it take the cold water to become completely blue? ____________

Answers *(Note: Unless students are to self-check their responses, fold the following answers under before reproducing the page.)*

1. This should only take a few seconds.
2. This should take several seconds to several minutes.

Make Your Own Thermometer

The thermometer you have in your house usually contains either colored alcohol or a special liquid metal known as *mercury.* Alcohol thermometers are usually red inside, while mercury thermometers look like they contain silver. However, both work the same way. A thermometer measures temperature by the expanding or contracting of liquid contents. Hot temperatures cause the liquid to expand and rise inside the thermometer. Cold temperatures cause the liquid to contract and fall inside of the thermometer.

You can make your own alcohol thermometer. Here is how.

Materials:

- pan
- cold water
- hammer
- clay
- hot water
- food dye
- ice
- nail
- rubbing alcohol
- clear drinking straw
- empty plastic bottle of drinking water with a screw-on lid

Procedure:

1. Use the hammer and the nail to make a hole in the center of the screw-on lid of the plastic bottle. (The hole should be just large enough to allow the straw to enter.)
2. Fill the bottle three-fourths of the way full with alcohol and add a few drops of food dye for color.
3. Insert the straw into the lid so that when the lid is screwed onto the bottle, the straw is about one inch (2.5 cm) below the surface of the alcohol.
4. Screw the lid onto the bottle. Make certain that it is tight.
5. Pack clay tightly around the lid opening to prevent air from entering or escaping from the bottle (not so tightly that it blocks the straw, however).
6. Place your thermometer in a pan. Add hot or cold water and observe the results.
7. Continue trying different temperatures of water in the pan.

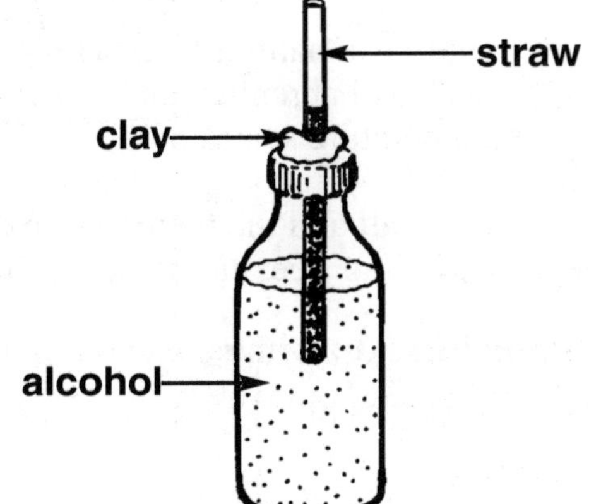

Results:

In hot water, the alcohol in the thermometer will expand and rise into the straw. In cold water, the alcohol will contract and will recede from the straw.

Closure:

1. Does your thermometer rise and fall to the same level as a classmate's thermometer? ______________________
2. How could you calibrate your thermometer? ______________________________________

____

Answers *(Note: Unless students are to self-check their responses, fold the following answers under before reproducing the page.)*

1. No, each thermometer will be unique.
2. Place the homemade thermometer in water next to a real thermometer in the same water. Use a permanent marking pen to mark the degrees on the homemade thermometer.

A Degree in History

The units for air temperature are usually recorded in either degrees Celsius or degrees Fahrenheit. The United States often uses the Fahrenheit scale, while many other countries use the Celsius scale to report their outside air temperatures. You can use the following formulas to convert between the degrees Celsius and Fahrenheit.

To convert Celsius to Fahrenheit: **F = (C x 9/5) + 32**

To convert Fahrenheit to Celsius: **C = (F - 32) x 5/9**

The History of the Thermometer:

The Fahrenheit scale was one of the first scales developed to measure temperature. It was developed by Gabriel Fahrenheit (1686–1736), a German physicist and meteorologist. Fahrenheit based his scale on two fixed points. The first point, zero, was set to the coldest temperature he could reach. To reach this temperature, he mixed salt with snow. His second point was 96 degrees, a temperature reached when his thermometer was held under somebody's tongue.

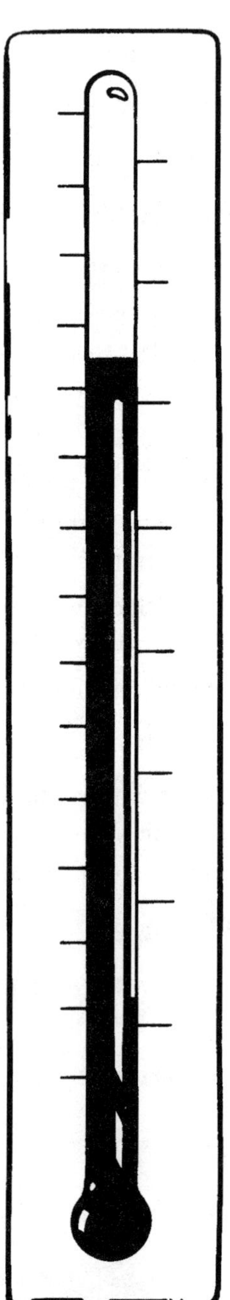

No one is sure why he decided the human body temperature should be 96 degrees (as opposed to an even 100). But, his scale is still popular today in countries such as the United States. Although today, most doctors agree the average body temperature is closer to 98.6 degrees rather than Fahrenheit's original estimate of 96 degrees.

The Celsius scale was developed by Swedish astronomer Anders Celsius (1701–1744). Celsius' scale is also based on two fixed points: the freezing point and boiling point of water (at sea level). Zero degrees Celsius is where water freezes and 100 degrees is where water boils (According to the Fahrenheit scale, water will freeze at 32 degrees and boil at 212). Most scientists use the Celsius scale when taking temperature measurements.

Temperature is a natural fact. Things are hot and things are cold. Temperature scales, such as Fahrenheit and Celsius, however, are created by people. Scientists create temperature scales based upon fixed reference points in order to label the temperatures we encounter. Fahrenheit, for example, used the temperature of ice mixed with salt and the temperature of the human body for his reference points. Celsius used the freezing point and melting point of water for his.

Extension Activity:

You can create your own temperature scale, using your homemade thermometer.

Place your thermometer in a freezer for 30 minutes and mark the level of the alcohol in the straw. This can be your zero temperature.

Place your thermometer in a pan of very hot water for one to two minutes and mark the level of alcohol in the straw. This can be your 100 degree temperature.

Closure:

Can you think of a place where your homemade thermometer would read less than zero? Where do you think it would read more than 100? Write your ideas on the back of this paper.

Temperature and Height

In this activity, you will be investigating temperatures at different heights around your home or school. The object of this investigation is to determine if there is any difference in the temperature closer to the ground as opposed to the temperature further above the ground. You will also be investigating how these temperatures change throughout the day.

Materials:

- aluminum foil
- five empty paper towel rolls
- five thermometers
- one tall stepladder (with at least five rungs)
- tape

Procedure:

1. Wrap the outside of each paper towel roll in a layer of aluminum foil.
2. Set up the tall stepladder outside in the sunlight.
3. Tape one paper towel roll, horizontally, to each rung of the ladder.
4. Insert a thermometer into each roll so that it is completely inside the roll.
5. Look at the chart on page 10 and take a temperature reading from each thermometer at the times indicated. Be sure to take the temperatures simultaneously (this may require five people) and do not touch the bulb of the thermometer with your hands (this could affect the results).

Results:

You should find a slight difference in temperature at each rung of the ladder. Generally, the higher thermometers will register a lower temperature.

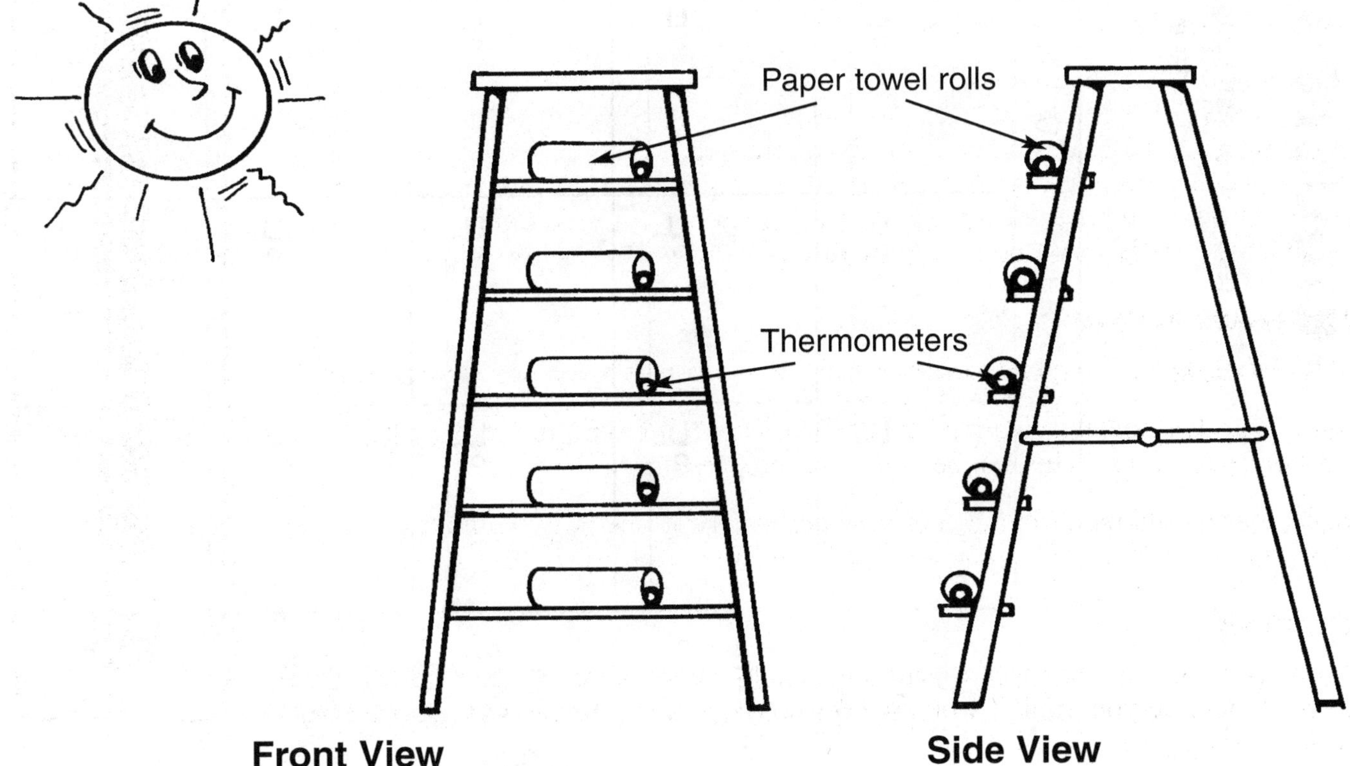

Temperature and Height *(cont.)*

Instructions: Follow the directions on page 9 for setting up this experiment. Record the temperature in degrees Fahrenheit every hour on the hour.

Height	8 A.M.	9 A.M.	10 A.M.	11 A.M.	12 P.M.	1 P.M.	2 P.M.	3 P.M.
5th Rung	_____°F	_____°F	_____°F	_____°F	_____°F	_____°F	_____°F	_____°F
4th Rung	_____°F	_____°F	_____°F	_____°F	_____°F	_____°F	_____°F	_____°F
3rd Rung	_____°F	_____°F	_____°F	_____°F	_____°F	_____°F	_____°F	_____°F
2nd Rung	_____°F	_____°F	_____°F	_____°F	_____°F	_____°F	_____°F	_____°F
1st Rung	_____°F	_____°F	_____°F	_____°F	_____°F	_____°F	_____°F	_____°F

Temperature and Height *(cont.)*

Graph

Instructions: Using the data from the chart on page 10, graph your results. This graph will contain five lines, one line for each rung of the ladder.

First, you must choose a color to represent each rung. Color in the small boxes in the legend with the color you have chosen to represent each rung.

Next, create a line graph by plotting the temperatures.

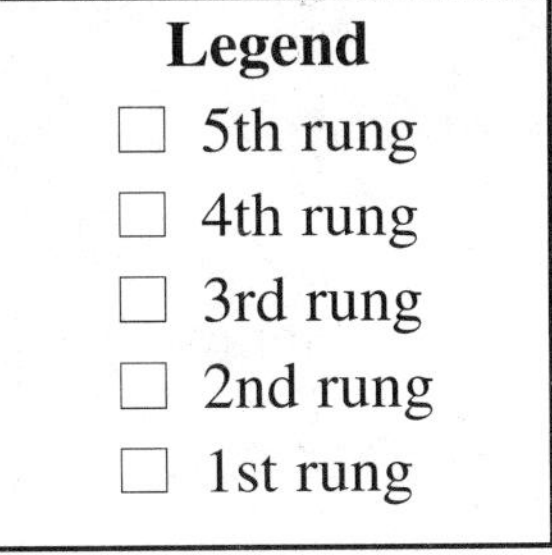

Legend

- ☐ 5th rung
- ☐ 4th rung
- ☐ 3rd rung
- ☐ 2nd rung
- ☐ 1st rung

90°
88°
86°
84°
82°
80°
78°
76°
74°
72°
70°
68°
66°
64°
62°
60°
Fahrenheit Temperature
8 A.M.
9 A.M.
10 A.M.
11 A.M.
12 P.M.
1 P.M.
2 P.M.
3 P.M.
Time

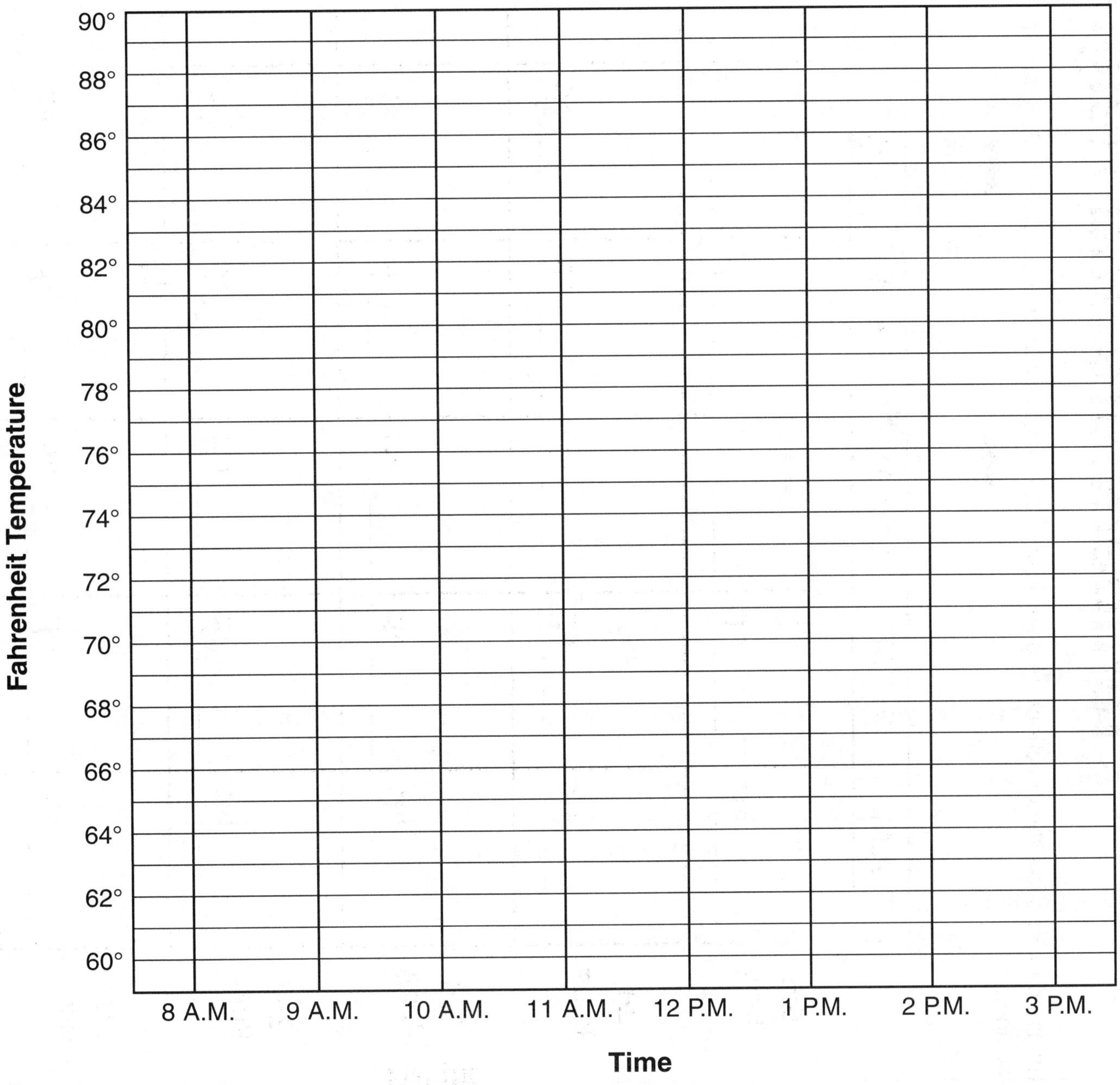

Temperatures Around a Building

In this activity, you will be determining the different temperatures which can be found around your home or school building during the day. Since the sun's position in the sky changes throughout the day, the different sides of a building will go through periods of being hotter and colder.

Materials:

- aluminum foil
- four empty paper towel rolls
- four thermometers
- duct tape

Procedure:

1. Wrap the outside of each paper towel roll in a layer of aluminum foil.
2. Tape one paper towel roll, horizontally, to each side of the building.
3. Insert a thermometer into each roll so that it is completely inside the roll.
4. Look at the chart on page 13 and take a temperature reading from each thermometer at the times indicated. Be sure to take the temperatures simultaneously (this may require four people) and do not touch the bulb of the thermometer with your hands (this could affect your results).

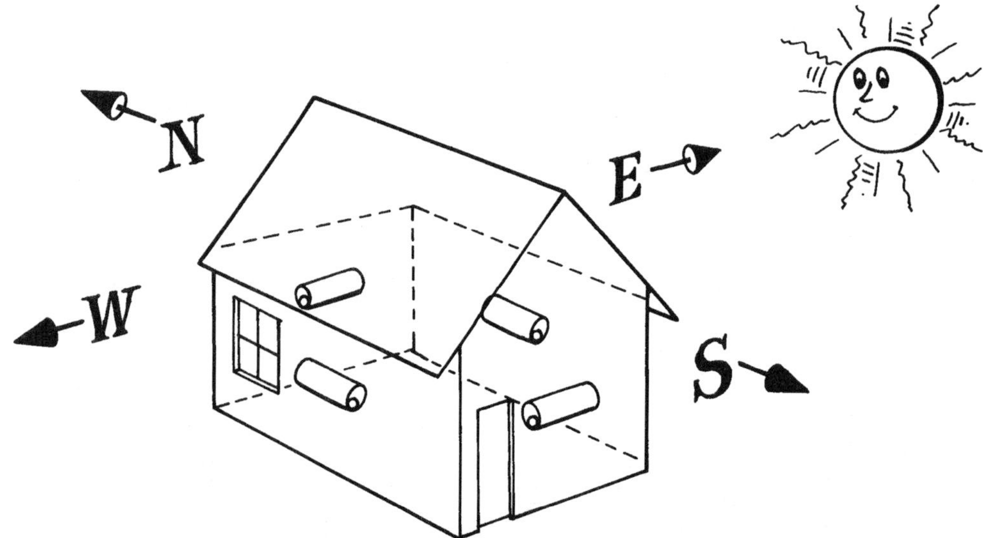

Results:

Throughout the day you should notice a marked change in each side of the building. For example, in the morning, the eastern side of the building will probably be warmer than the western side. In the afternoon, the western side should be warmer.

Internet Extender

About Temperature
http://www.unidata.ucar.edu/staff/blynds/tmp.html

Summary: This Web site has great information, including a description of temperature and history of the development of the thermometers and temperature scales. This is especially designed for students in grades 5-8.

Temperatures Around a Building *(cont.)*

Chart

Instructions: Follow the directions on page 12 for setting up this experiment. Record the temperature in degrees Fahrenheit every hour, on the hour.

Side of Building	8 A.M.	9 A.M.	10 A.M.	11 A.M.	12 P.M.	1 P.M.	2 P.M.	3 P.M.
North	______°F	______°F	______°F	______°F	______°F	______°F	______°F	______°F
South	______°F	______°F	______°F	______°F	______°F	______°F	______°F	______°F
East	______°F	______°F	______°F	______°F	______°F	______°F	______°F	______°F
West	______°F	______°F	______°F	______°F	______°F	______°F	______°F	______°F

Temperatures Around a Building

(cont.)

Graph

Instructions: Using the data from the chart on page 13, graph your results. This graph will contain four lines, one line for each side of the building.

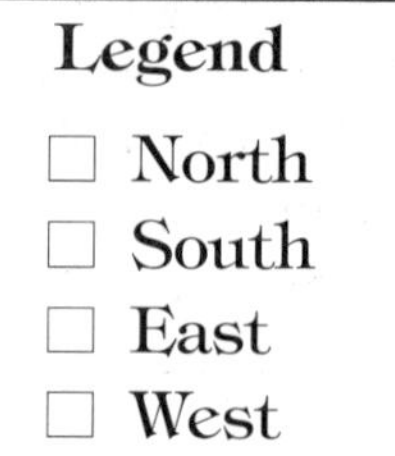

First, you must choose a color to represent each side of the building. Color the small boxes in the legend with the color you have chosen to represent each side of the building.

Next, create a line graph by plotting temperatures.

Legend

- ☐ North
- ☐ South
- ☐ East
- ☐ West

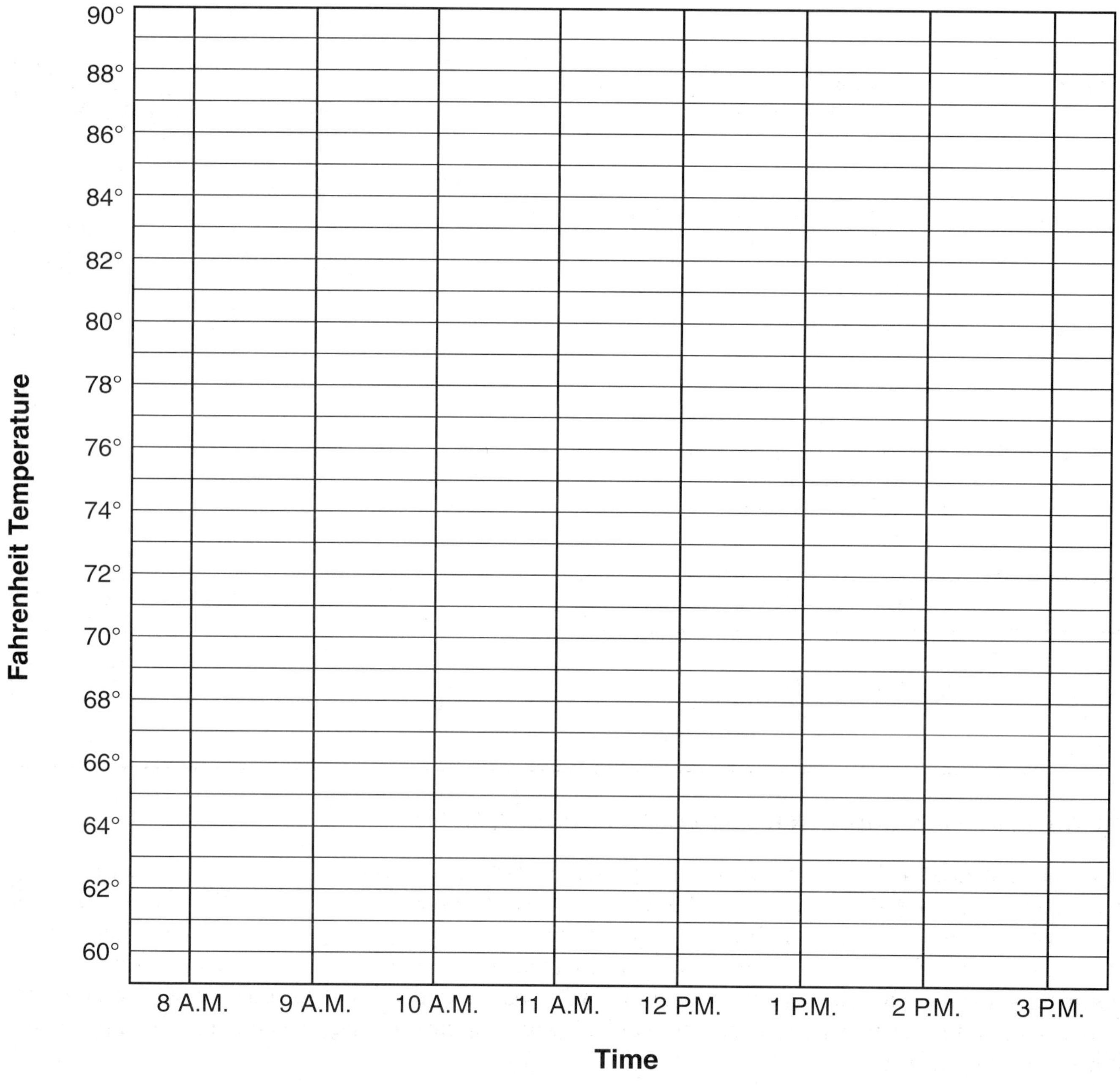

90°
88°
86°
84°
82°
80°
78°
76°
74°
72°
70°
68°
66°
64°
62°
60°
Fahrenheit Temperature
8 A.M.
9 A.M.
10 A.M.
11 A.M.
12 P.M.
1 P.M.
2 P.M.
3 P.M.
Time

Air Pressure and Elevation

Ms. Frizzle's class not only studies temperature before embarking on their trip, they also study the atmosphere and air pressure. They must learn about air if they are to understand the weather. You can see some more details about our atmosphere on page 16.

As you can see on page 16, there is an awful lot of atmosphere above your head. In fact, every square inch of your body is pressed down upon by 14.7 pounds of air pressure (at sea level) at all times. You do not feel it because you are used to it. As you climb a mountain or ride in an airplane, the pressure becomes less since you are moving higher into the atmosphere and less air is above your head. Look at the picture below to see how the air pressure decreases as you climb higher into the atmosphere.

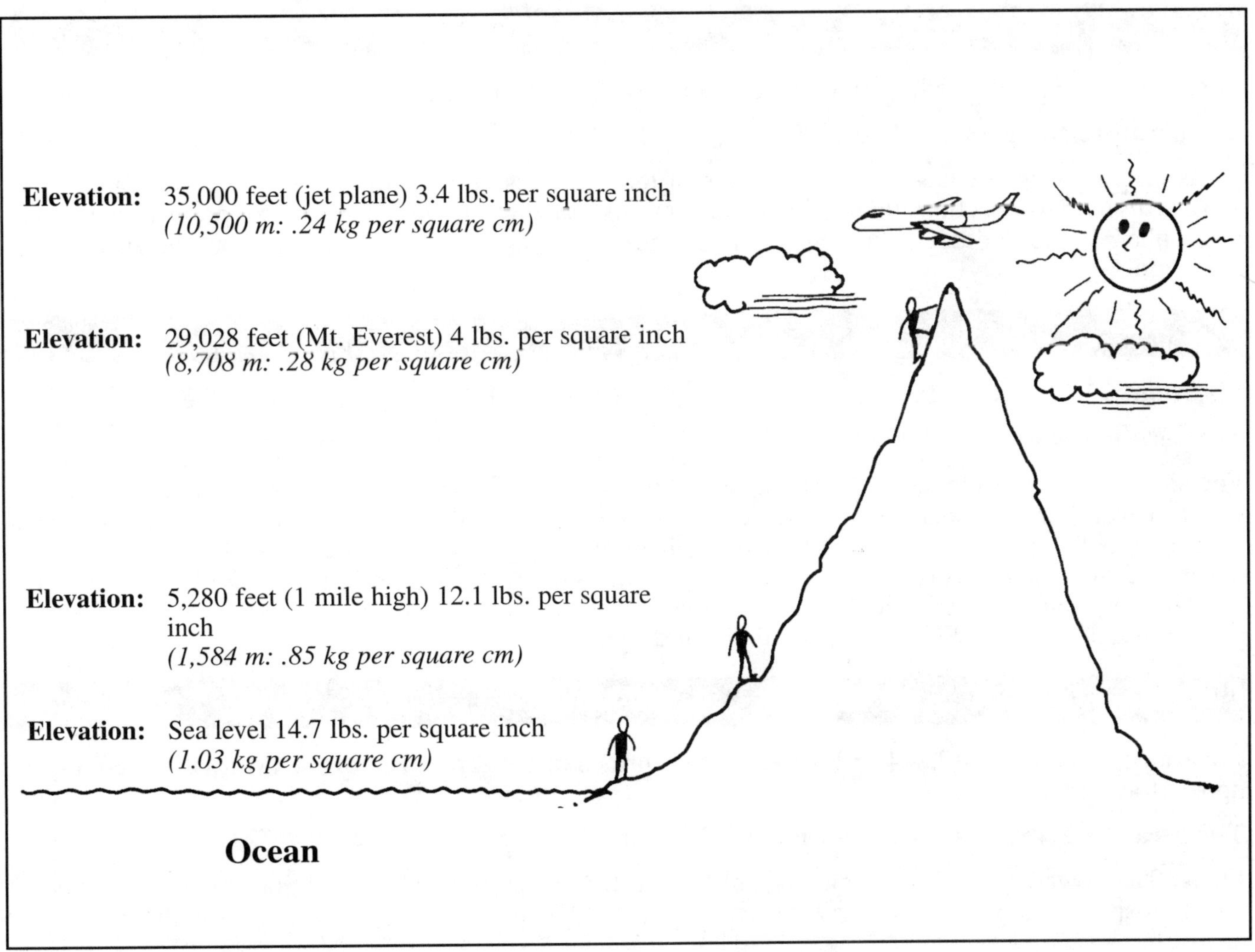

Why do your ears pop when you go up in an airplane?

You are not crushed by the 14.7 pounds of air pressure pushing on every square inch (1.03 kg per square centimeter) each day because you have internal pressure pushing out against the atmosphere at an equal rate. When you climb high into the atmosphere, the air pressure is greatly reduced, and the pressure in your body fights to get out to become equal again. A tube leading from your inner ear to your throat, known as the *Eustachian* tube, allows air to pass into the throat to equalize the pressure. If you have a cold, the Eustachian tube can be blocked, and your ears will hurt since the air is trapped and will press on the eardrum.

Atmospheric Layers

Thermosphere

Location: 50–70 miles (80.5–112.7 km) above the surface of the earth

Temperature Range: 2,192°F to 4°F (1,200°C to 20°C)

Facts: The thermosphere, also known as the ionosphere, is the highest layer of the atmosphere. Beyond the thermosphere lies the void of outer space. The aurora borealis, a spectacular electrical display, occurs in the thermosphere. Also, radio waves can be reflected by the thermosphere to carry signals farther.

Mesosphere

Location: 30–50 miles (48.3–80.5 km) above the surface of the earth

Temperature Range: –112°F to 68°F (–80°C to 20°C)

Facts: The mesosphere has the greatest range of temperatures in the entire atmosphere. It is generally warmer in its lower sections than in its upper sections. The aurora borealis also can occur in the upper mesosphere. High altitude helium weather balloons can reach into the mesosphere, but the air is too thin for airplanes to fly.

Ozone Layer

Location: 12–30 miles (19.32–48.3 km) above the surface of the earth (It is a part of the stratosphere.)

Temperature Range: 32°F to 68°F (0°C to 20°C)

Facts: Ozone is a poisonous gas for humans to breathe, but its presence in the stratosphere is important for all life on Earth. Ozone blocks harmful ultraviolet rays from the sun, which cause sunburns and can alter the DNA in plants and animals to cause genetic mutations. Recently, scientists have become concerned that chemicals known as chlorofluorocarbons (CFCs), found in aerosol sprays, refrigerators, and air conditioners, may be depleting the ozone layer. As a result, world governments have agreed to stop production of all CFC products by the year 2000.

Stratosphere

Location: 8–30 miles (12.8–48.3 km) above the surface of the earth (The ozone layer is located in the upper stratosphere.)

Temperature Range: –67°F to –40°F (–55°C to –40°C)

Facts: The powerful winds of the jet stream are located in the stratosphere. Major airline planes fly in the stratosphere. Since hardly any water exists in the stratosphere, there are no clouds at this level or any of the levels above it.

Troposphere

Location: 0–8 miles (0–12.88 km) above the surface of the earth

Temperature Range: –128°F to 136°F (–89°C to 58°C) near the earth's surface; –67°F to 68°F (–55°C to 20°C) in the upper troposphere

Facts: All of the earth's weather occurs in the troposphere. Non–jet airplanes fly in the troposphere. As a general rule, the temperature of the troposphere will decrease 3°F (5.5°C) for every 1,000 feet (300 m) of elevation.

Air Pressure Demonstration

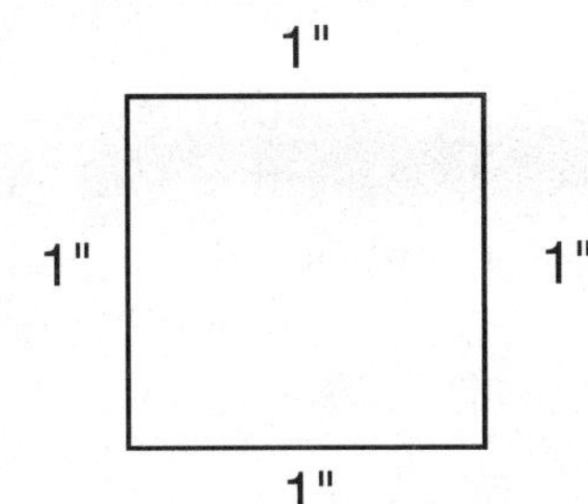

This demonstration will illustrate the amount of air which sits above your head every day. The weight of the air is called air pressure, and it is measured in units of pounds per square inch. At sea level, there are approximately 14.7 pounds of air on every square inch (1.03 kg per square centimeter) of the ground.

Materials:

- a full newspaper sheet
- one long and thin stick (A paint stirrer works well.)
- table

Procedure:

1. Spread the full sheet of newspaper on the table.
2. Insert the stick under the sheet of newspaper so that 1/2 of the stick is projecting over the edge of the table.
3. Using your fist, in a quick motion, strike down on the exposed end of the stick.

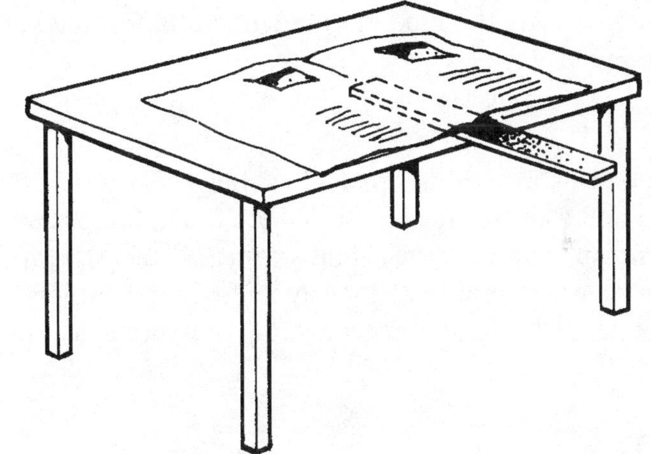

Results:

If struck quickly enough, the stick will break. The large surface area of the newspaper will distribute the air pressure and make the paper incredibly heavy. An average sheet of newspaper has close to 10,000 pounds (4,500 kg) of air pressure on it!

Closure:

Why is it not difficult to move the newspaper if you strike the stick slowly?____________________

Answer *(Note: Unless students are to self-check their responses, fold the following answers under before reproducing the page.)*

Slowly striking the stick allows the air molecules to roll off of the paper as if they were a bunch of marbles. Striking it firmly does not give the air molecules a chance to get out of the way.

Barometers and Air Pressure

As the class departs for its field trip, they encounter a burst of strong wind. Arnold reports that wind is caused when a mass of heavier air pushes out a mass of lighter air. The mass of heavier air has a greater air pressure than the mass of lighter air. Meteorologists measure the air pressure to predict weather changes and wind patterns.

Meteorologists have been measuring air pressure since 1643 when Evangelista Torricelli invented a device known as a *barometer.* A barometer is composed of a thin glass tube first filled with mercury and then placed upside down in a dish of mercury. Air pressure pushes down on the dish of exposed mercury and this causes the mercury in the tube to rise. If the air pressure drops, so will the level of the mercury in the tube. In the United States, the barometer is read in inches of mercury.

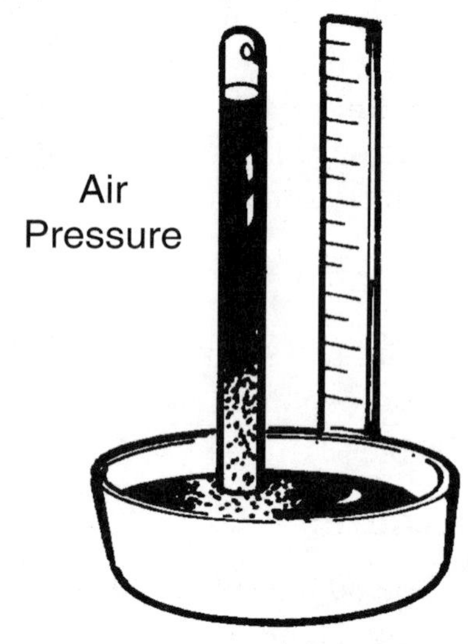

Barometer

As the earth spins on its axis and is alternately warmed and cooled by the action of day and night, areas of high and low air pressure are created. High pressure areas drive out clouds and storms while low pressure areas attract clouds and storms. If the mercury level in the barometer is rising, that means a high pressure system is on its way and better weather can be expected. If the mercury level in the barometer is falling, that means a low pressure system is on its way, and a storm may be moving into your area.

Activity:

Watch the local weather report tonight on the news or read about the weather in the newspaper.

1. Is the barometric pressure rising or falling?

 ____

2. If it is rising, what kind of weather is the weather person predicting?

 ____

 ____

 ____

3. If it is falling, is the weather person predicting a storm?

 ____

 ____

 ____

The Beaufort Scale

Arnold notices that the wind is blowing hard enough to blow the leaves he has been raking. Winds can have a variety of speeds, and these different speeds can cause different effects on land and at sea. In 1805, a scale was devised to describe these different wind speeds and their levels of strength. The *Beaufort Wind Scale,* developed by British rear admiral Sir Francis Beaufort, provided sea captains with information about the winds which would affect their sailing. Below is a copy of his scale and the symbols he used to represent each level.

Beaufort Wind Scale

Beaufort Number	Speed mph	Speed kph	Wind Name	Land Indication	Symbol
0	less than 1	less than 1	calm	smoke rises vertically	
1	1–3	1–5	light air	smoke drifts with wind	
2	4–7	6–11	light breeze	leaves rustle, weather vanes move	
3	8–12	12–19	gentle breeze	twigs move, flags extend	
4	13–18	20–29	moderate breeze	small branches move, dust rises	
5	19–24	30–38	fresh breeze	small trees sway	
6	25–31	39–50	strong breeze	large branches sway, wind howls	
7	32–38	51–61	moderate gale	whole trees move, walking is hard	
8	39–46	62–74	fresh gale	twigs and branches break from trees	
9	47–54	75–86	strong gale	branches break, roof tiles fly off	
10	55–63	87–101	whole gale	trees uprooted, structural damage	
11	64–72	102–116	storm	widespread damage	
12	above 72	above 116	hurricane	much destruction and damage	

Building an Anemometer

The Beaufort scale describes what winds of different speeds do. We can use this scale to help us calibrate an instrument for determining the speeds of the winds. This instrument is called an *anemometer.* An anemometer records the wind speed in *revolutions per minute (RPM).* RPM is the number of times the anemometer spins around in one minute. The faster the wind blows, the higher the RPMs.

Materials:

- $2\frac{1}{2}$ inch (6.35 cm) Styrofoam ball
- two ping-pong balls
- two meat skewers
- long knitting needle
- plastic drinking straw
- coffee can (with lid)
- sand
- permanent marking pen
- tape
- craft knife

Procedure:

1. Carefully cut the two ping-pong balls in half. (Your teacher may wish to do this step and step 5 ahead of time, using a craft knife.) This will create a total of four hemispheres. Color one of the hemispheres with the permanent marking pen.
2. Run one of the meat skewers through the middle of the Styrofoam ball so that it projects equally out of both sides of the ball.
3. Run the other meat skewer through the middle of the Styrofoam ball so that it is perpendicular to the first.
4. Your meat skewers should form a plus sign if steps 2 and 3 are done correctly.
5. Poke two holes in each of the four ping-pong hemispheres. The holes should be made close to the edge of each ball and line up across the diameters.
6. Insert the four ends of the meat skewers into the four halves of the ping-pong balls. Each half should have its hollow side facing the same direction.
7. You might want to use a small amount of tape or glue to secure the ping-pong balls to the skewers.
8. Drive the knitting needle straight through the top of the Styrofoam ball so that it intersects the plane of the meat skewers at a right angle.
9. Cut the straw a little longer than the height of the coffee can.
10. Use the sharp end of the knitting needle to then poke a hole in the coffee can lid. Make the hole large enough to fit the straw through.
11. Insert the end of the knitting needle into the drinking straw.
12. Insert the drinking straw, with the needle in it, through the hole in the lid.
13. Fill the coffee can with sand so that the drinking straw will stand up straight.
14. Place the lid on the coffee can, and you are ready to go!

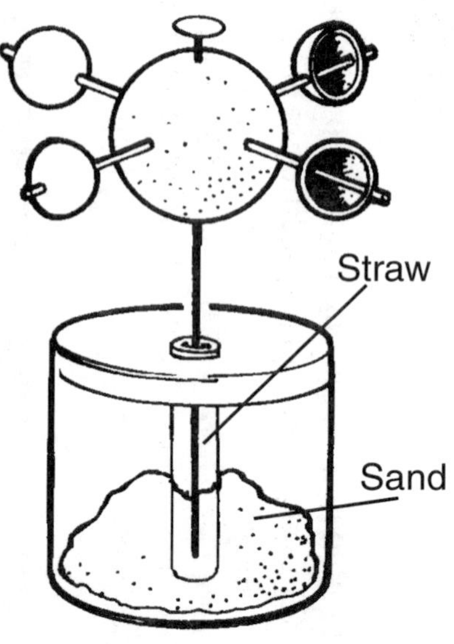

Using Your Anemometer

Instructions: Use your anemometer to record the wind speed (in RPMs) near your home or school. One location which might work well for this activity is the playground. Can you think of other places to record wind speed?

Count the number of revolutions your anemometer makes in 30 seconds and multiply your answer by two to find the number of revolutions in one minute. Use the Beaufort scale on page 19 to fill in the columns for the wind name and land indication based on your observations of the wind.

Location	RPMs	Wind Name	Land Indication
Playground			

Building a Weather Vane

An anemometer is helpful in determining a wind's speed. In order to determine a wind's direction, you will need to build a *weather vane.* A weather vane is a device which changes its direction according to how the wind blows. Follow the directions below to make your own weather vane, and then perform the activity which follows.

Materials:

- $2\frac{1}{2}$-inch (6.35 cm) Styrofoam ball
- meat skewer
- long, thin knitting needle
- plastic drinking straw
- coffee can (with lid)
- sand
- poster board
- tape
- magnetic compass
- scissors

Procedure:

1. Use the patterns on page 23 to trace and cut out a weather vane from your poster board.
2. Run the meat skewer through the middle of the Styrofoam ball so that it projects equally out of both sides of the ball.
3. Drive the knitting needle straight through the top of the Styrofoam ball so that it intersects the meat skewer at a right angle.
4. Cut the straw a little longer than the height of the coffee can.
5. Use the sharp end of the knitting needle to then poke a hole in the coffee can lid. Make the hole large enough to fit the straw through.
6. Insert the knitting needle and drinking straw into the coffee can lid.
7. Fill the coffee can with sand so that the drinking straw will stand up straight.
8. Place the lid on the coffee can.
9. Affix the arrowhead to one side of the meat skewer with tape.
10. Affix the tail to the other side of the meat skewer with tape.
11. Set your weather vane outside (or in front of a fan) and use your magnetic compass to determine which way the wind is blowing. The arrow will point into the direction from which the wind comes.

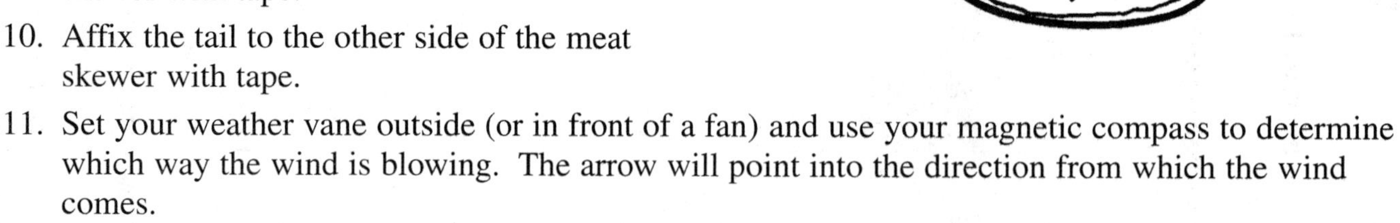

Internet Extender

Wind

http://www.whnt19.com/kidwx/wind.htm

Summary: Information at this Web site describes wind direction, the jet stream, and the anemometer.

Weather Vane Patterns

Use these patterns to create an arrowhead and tail on a piece of poster board. Then, follow the directions for making a weather vane on page 22.

arrowhead

tail

Using Your Weather Vane

Instructions: Use your weather vane to determine the directions of the winds near your home or school. You will need to use a magnetic compass and a watch in order to perform this activity.

Place your weather vane outside. Write down the direction of the wind on the line provided and the time at which you recorded the wind in the parentheses next to the line.

For example, if you find the wind direction for the playground to be north at 9:00 A.M., write N on the line (under Direction #1) and 9:00 A.M. in the parentheses. If you return to the playground at 10:30 A.M. and find the wind direction to be northwest, write NW on the line (under Direction #2) and 10:30 A.M. in the parentheses.

The playground is only a suggestion. What other areas might be interesting to test?

LOCATION	DIRECTION 1	DIRECTION 2	DIRECTION 3	DIRECTION 4
PLAYGROUND	____(________)	____(________)	____(________)	____(________)
________	____(________)	____(________)	____(________)	____(________)
________	____(________)	____(________)	____(________)	____(________)
________	____(________)	____(________)	____(________)	____(________)
________	____(________)	____(________)	____(________)	____(________)
________	____(________)	____(________)	____(________)	____(________)

Investigating Density of Air

After being blown by the wind, the magic bus mysteriously turns into a hot air balloon, and the class is whisked away into the sky. To explain this, Molly, Alex, and Rachel describe the ability of hot air to expand and rise while Carlos illustrates hot air's ability to float on cold air.

This phenomenon is known as *density.* If an object is dense, it will sink, as a rock sinks in water. If an object is not very dense, it will float, like a marshmallow in a cup of hot chocolate.

Hot objects are typically less dense than cold objects. Hot air is no exception, and since it is less dense than cold air, it will rise. You can witness how air rises firsthand with the following activity.

Materials:

- two glass bottles of equal size and shape
- red food dye
- warm water
- cold water
- 3 x 5 in. (7.6 x 12.7 cm) index card

Procedure:

1. Fill one bottle all the way to the brim with warm water.
2. Add several drops of red food dye to the warm water and stir.
3. Completely fill the other bottle with cold water.
4. Place an index card over the top of the cold water bottle and hold it in place with your hand.
5. Quickly invert the cold water bottle over the warm water bottle and line up the necks of the bottles. (Do this over a sink. It could get messy.)
6. The bottles should now be touching at their necks with the card in between them.
7. Carefully, but quickly, remove the card by pulling it towards you. Have a friend hold the bottles together as you remove the card.

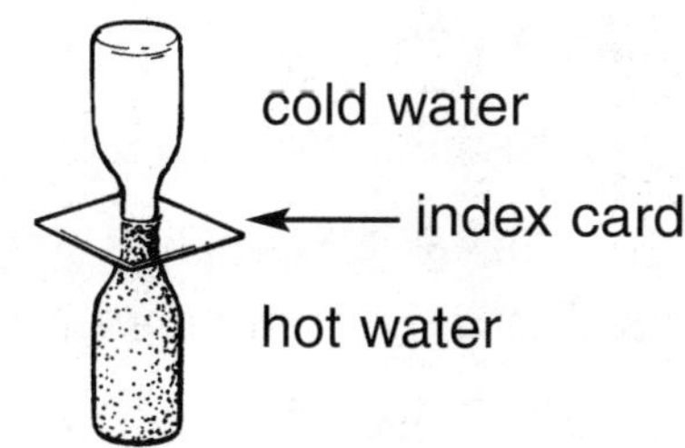

Results:

You should notice the red food dye rising upwards into the cooler water above. The warm water is lighter (less dense) than the cold water and will rise.

Closure:

Answer the following questions on the back of this paper.

1. What would happen if we tried this activity again, only this time the colored warm water was placed on the top instead of the bottom?
2. What would happen if we tried this activity again, only this time we colored the cold water and put it on the bottom?
3. What would happen if we tried this activity again, only this time we colored the cold water and put it on the top?

Answers *(Note: Unless students are to self-check their responses, fold the following answers under before reproducing the page.)*

1. The colored warm water would remain on the top.
2. The colored cold water would remain on the bottom.
3. The clear, warm water would rise into the colored cold water on top.

Dew Point

As the class continues to rise in the hot air balloon, Ms. Frizzle explains to the students that the warm air rising from the surface of the earth carries with it a great deal of water vapor. *Humidity* is the level of water vapor present in the air. This water vapor is then used to make the components of weather, such as clouds, rain, snow, and hail. You can determine the amount of water vapor in the air around you by conducting this simple experiment.

What is the dew point?

The *dew point* is the temperature at which water vapor in the air will become cold enough to turn from a gas to a liquid. A mirror or a window is usually as cold or cooler than the dew point of the water in your breath. Try breathing on a mirror to see dew formation in action and then try the following activity.

Materials:

- empty coffee can
- thermometer
- ice
- water
- food dye

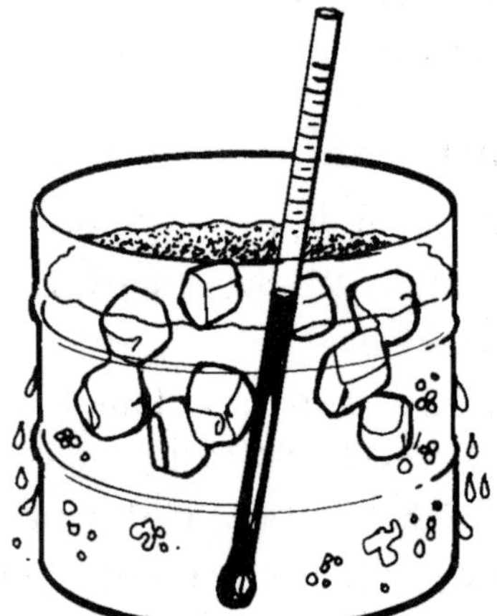

Procedure:

1. Fill the coffee can half way to the top with water.
2. Place a few drops of food dye in the coffee can.
3. Place the thermometer in the coffee can.
4. Begin to add ice slowly to the coffee can.
5. Watch the outside of the coffee can for signs of "sweating." This sweating is called condensation.
6. Immediately record the temperature when you see condensation. This is the dew point.

Closure:

Answer the following questions on the back of this paper.

1. Did the condensation which formed on the outside of the can come through the coffee can?
2. How do you know your answer to number 1 is correct?
3. Where else have you seen a phenomenon similar to this?
4. Would you expect to find dew on the outside of a hot cup of cocoa or coffee? Why or why not?
5. Why does the bathroom mirror sometimes fog up when you take a hot shower or bath?
6. Why do cold water baths or showers not cause dew to form on the mirror?

Answers *(Note: Unless students are to self-check their responses, fold the following answers under before reproducing the page.)*

1. No, the condensation came from the air around the can.
2. If the condensation came from the can, it would be colored.
3. An example might be the fog forming on a bathroom mirror.
4. No. The hot cup does not allow the water vapor in the air to cool enough to condense back into a liquid.
5. The mirror is cold enough to reach the dew point of the water vapor caused by the hot shower.
6. Hot air can hold more water vapor than cold air. A cold water shower will not heat the bathroom up enough for the dew point to be reached. Therefore, the dew will not form on the mirror.

Build a Sling Psychrometer

A weather term known as *relative humidity,* expresses the amount of water which is actually present in the air as a percent of the amount of water which could be in the air. You can determine the relative humidity in your area, using a device known as a *sling psychrometer*. A sling psychrometer is composed of two thermometers. One of the thermometer's bulbs is covered in a wet cloth and is called a wet bulb thermometer. The dampness causes that thermometer to give a cooler reading than the actual temperature. The difference in temperature between the two thermometers can then be interpreted, using a table, to determine the percentage of relative humidity.

Materials:

- drill
- small block of wood
- old shoelace
- four u-shaped clamps
- short stick
- thick string
- screws
- copy of page 28
- two small alcohol thermometers

Procedure:

1. Ask your teacher, or another helpful adult, to drill a small hole near the top (and towards the center) of the block of wood and at the end of the short stick. This is where the string will be threaded.
2. Using the u-shaped clamps and screws, secure the thermometers to the small block of wood. The thermometers should be parallel to each other and their bulbs pointed away from the hole. Be sure to allow their bulbs to project off the end of the block of wood.
3. Make the wet bulb thermometer by cutting a small length of shoelace so that it will cover one of the thermometer's bulbs. Slip the shoelace over one of the bulbs and secure it into place by tying a string around it so it will not slip off.
4. Cut an 8-inch (20.32 cm) length of string, tie one end to the hole in the short stick, and tie the other to the hole in the small block of wood. Tie both ends of the string tightly.

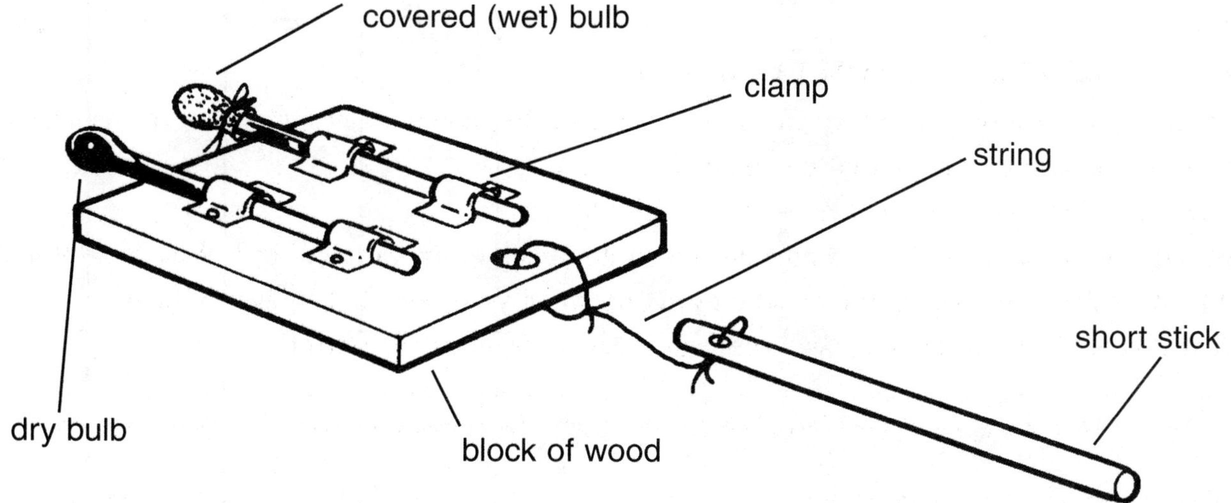

Internet Extender

Humidity and Dew Point

http://www.whnt19.com/kidwx/humidity.html

Summary: This Web site has a good explanation of humidity and dew point.

Build a Sling Psychrometer *(cont.)*

Using Sling Psychrometer Results to Determine Relative Humidity

Instructions: Use your sling psychrometer to determine the relative humidity in the air. First, you will need to dip the bulb of the shoelace-covered thermometer in some water. Next, find a clear and open space outside. Make sure that there are no other students near you. Then, lightly swirl your psychrometer in the air for about 30 seconds. Repeat this five more times. Record the temperatures of the wet and dry bulbs below. For each swirl, subtract the temperature of the wet bulb from that of the dry bulb to find the difference.

Use the table on page 29 to determine the relative humidity and record it below. Indicate the time of day you took the measurement so that you can compare relative humidities throughout the day.

Dry Bulb (°F)	Wet Bulb (°F)	Difference	Relative Humidity (%)	Time
°F	°F		%	
°F	°F		%	
°F	°F		%	
°F	°F		%	
°F	°F		%	
°F	°F		%	

Build a Sling Psychrometer *(cont.)*

Percent of Relative Humidity Table

Instructions: Use this table to find the relative humidity of the air. The temperature of the dry bulb runs down the left side of the table, while the difference in temperature between wet and dry bulbs runs along the top. Place one finger on the temperature of the dry bulb and another finger on the difference between the wet and dry bulbs. Run your fingers down and across the column and row. The number where your fingers meet indicates the percentage of relative humidity.

Percent of Relative Humidity (°F)

Degree Difference Between the Dry and Wet Bulbs

Temperature of Dry Bulb	1	2	3	4	5	6	7	8	9	10	11	12	13	14	15	16	17	18	19
98	96	93	89	86	82	79	96	72	69	66	63	60	57	54	52	49	46	44	41
96	96	93	89	85	82	78	75	72	68	65	62	59	57	54	51	48	45	43	40
94	96	93	89	85	81	78	75	71	68	65	62	59	56	53	50	47	44	42	39
92	96	92	88	85	81	78	74	71	67	64	61	58	55	52	49	46	43	40	38
90	96	92	88	84	81	77	74	70	67	63	60	57	54	51	48	45	42	39	36
88	96	92	88	84	80	77	73	69	66	63	59	56	53	50	47	44	41	38	35
86	96	92	88	84	80	76	72	69	65	62	58	55	52	49	45	42	39	36	33
84	96	92	87	83	79	76	72	68	64	61	57	54	51	47	44	41	38	35	32
82	96	91	87	83	79	75	71	67	64	60	56	53	49	46	43	40	36	33	30
80	96	91	83	87	79	74	70	66	63	59	55	52	48	45	41	38	35	31	28
78	95	91	86	82	78	74	70	66	62	58	54	50	47	43	40	36	33	30	26
76	95	91	86	82	78	73	69	65	61	57	53	49	45	42	38	34	31	28	24
74	95	90	86	81	77	72	68	64	60	56	52	48	44	40	36	33	29	26	22
72	95	90	85	80	76	71	67	63	58	54	50	46	42	38	34	31	27	23	20
70	95	90	85	80	75	71	66	62	57	53	49	44	40	36	32	28	24	21	17
68	95	90	84	79	75	70	65	60	56	51	47	43	38	34	30	26	22	18	15
66	95	89	84	78	74	69	64	59	54	50	45	41	36	32	28	23	20	16	12
64	94	88	83	77	73	68	63	58	53	48	43	39	34	30	25	21	17	13	9
62	94	88	83	77	72	67	61	56	51	46	41	37	32	27	23	18	14	10	5
60	94	88	82	77	71	65	60	55	50	44	39	34	29	25	20	15	11	6	2

Cloud Formation

The meteorologists at the weather station are wondering when Ms. Frizzle will be bringing her class for their field trip. Little do they know that the class is high above in a balloon which is drifting into a large cloud formation. The class is learning firsthand how clouds are formed.

Cloud Formation in a Nutshell

1. Warm, moist air rises into the upper atmosphere.
2. The upper atmosphere has lower air pressure than the lower atmosphere.
3. Under the conditions of lower air pressure, the warm air expands, cools, and forms clouds.

All clouds form under these three conditions. But warm, moist air begins the process of cloud formation in one of the three following ways:

1. *Convection*—When the sun warms the ground, the air above the ground is heated and begins to rise. This warm air, containing water vapor, rises in a process known as convection.

   Since the upper atmosphere has less pressure than the lower atmosphere, the rising air and water vapor will expand and become cooler. The cooling water vapor forms a fog. Fogs which are high above the ground are called clouds.
2. *Lifting*—Clouds are formed by lifting when warm, moist air moves up the side of a mountain. Along its journey, the air is lifted higher into the atmosphere where the pressure is lower. Again, the water vapor in the air is allowed to cool and form a fog. This is why we often see clouds over mountains.
3. *Frontal Activity*—Sometimes a mass of warm, moist air, known as a warm front, will run into a mass of colder air, known as a cold front. When the two meet, the warm, moist air rises higher into the atmosphere. The lower pressure of the upper atmosphere causes the water vapor in the warm air to cool and form a cloud.

Internet Extender

Cloud Catalog

http://covis.atmos.uiuc.edu/guide/clouds/

Summary: Topics covered by this Web site include clouds at high, middle, and low levels; how clouds form; convective clouds; and other types of clouds.

Cloud Formation *(cont.)*

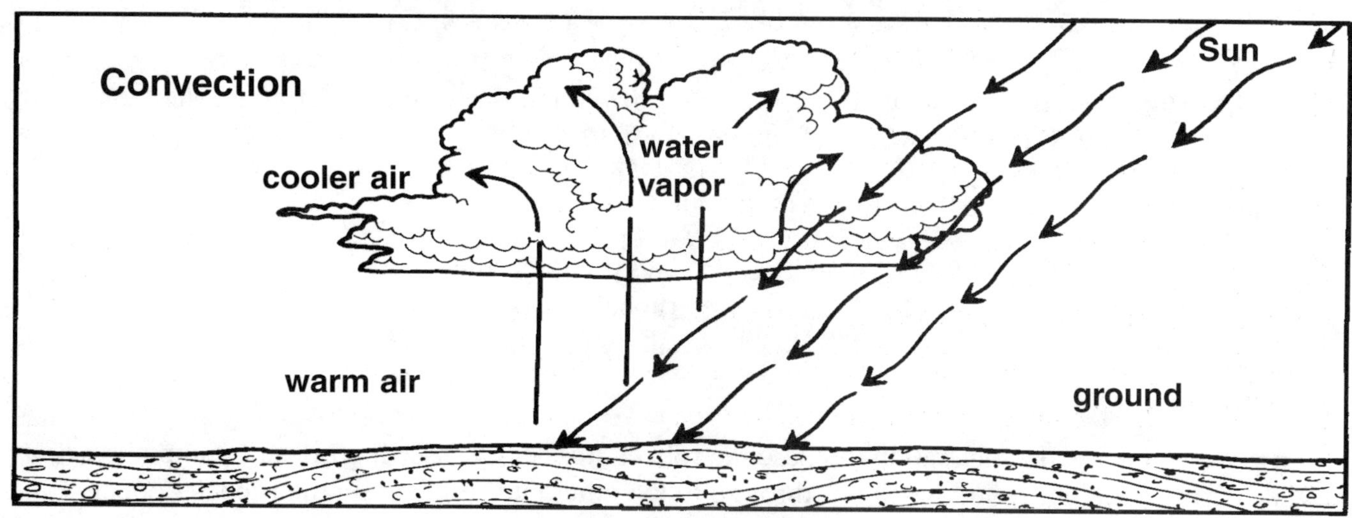

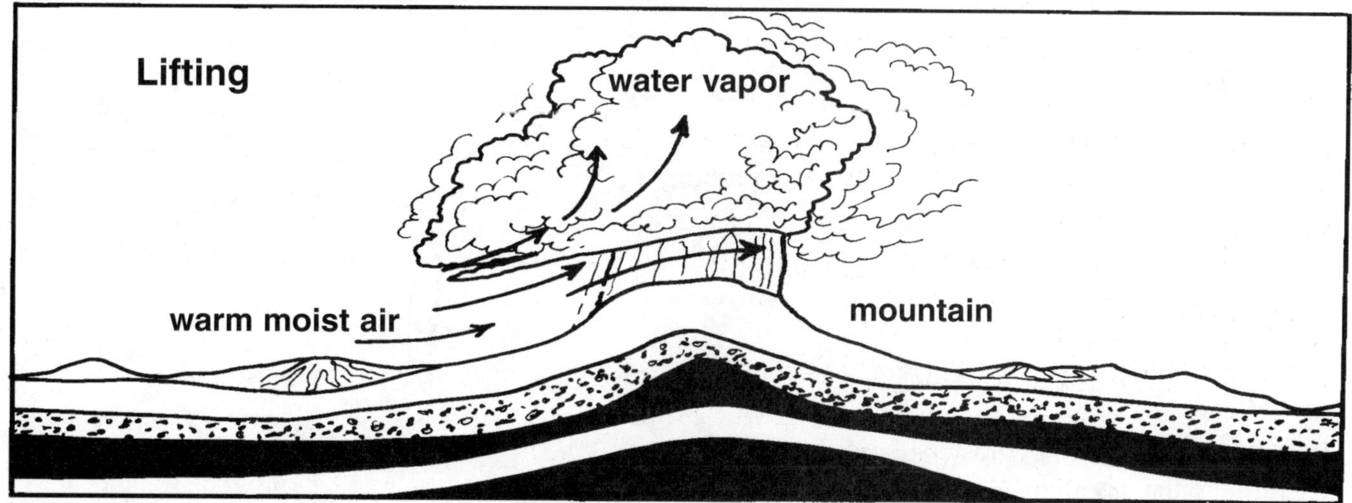

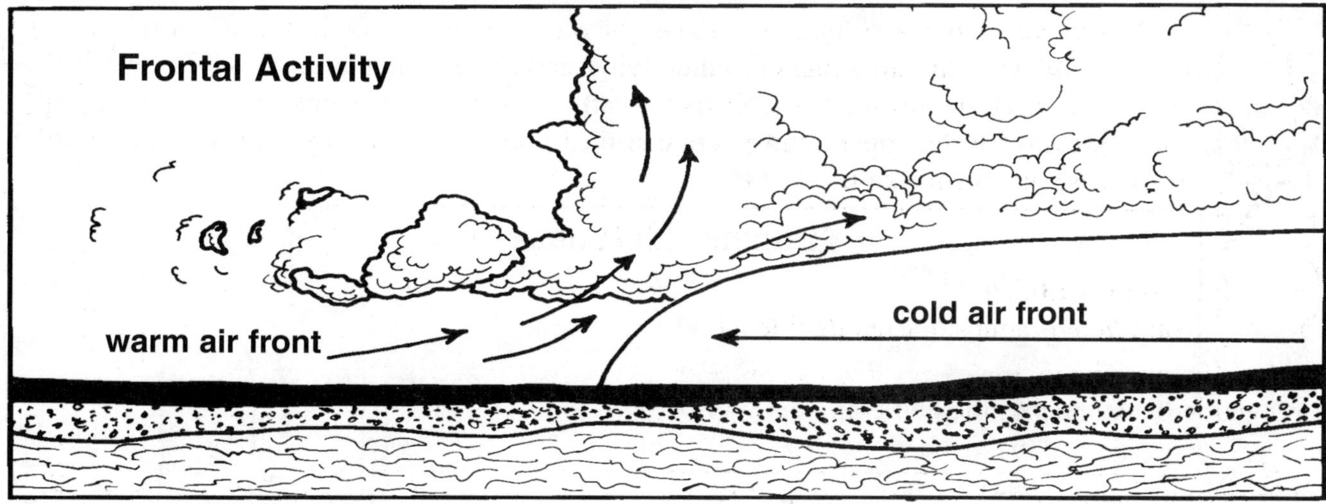

Cloud in a Bottle— Teacher Demonstration

Teacher Instructions: You can simulate the effects of the conditions of the upper atmosphere on warm, moist air with a demonstration known as the cloud in a bottle.

Materials:

- a 2-liter plastic soda bottle
- rubbing alcohol (or water)
- bicycle pump
- ball pump needle
- rubber stopper
- safety goggles

Preparation:

Before performing this demonstration, prepare the rubber stopper by inserting the ball pump needle into the wide end of the stopper and passing it through to the other end. Be certain the rubber stopper will fit snugly into the mouth of the plastic soda bottle. **Note:** Goggles should be worn by you and your helper as a precaution.

Procedure:

1. Pour a tablespoon of rubbing alcohol into the bottle (water will also work, but not as well as rubbing alcohol).
2. Place the stopper on the bottle and attach the bicycle pump to the needle.
3. Have a helper hold the stopper in place as you pump.
4. After three to five pumps, ask your helper to remove the stopper.

Results:

A white fluffy cloud will form in the bottle. Essentially, you have simulated the behavior of water vapor in the upper atmosphere. As the pressure in the bottle rapidly decreased due to the rapid escape of air, the alcohol vapor (or water vapor) was forced to quickly expand. This rapid expansion caused the vapor to cool quickly and form cloud droplets.

Alcohol will form a cloud with less of a pressure reduction than will water; therefore, alcohol is recommended, but not essential, for this demonstration.

Extension:

By replacing the stopper in the bottle and applying pressure, see if you can make the cloud disappear. This demonstration can be repeated many times.

alcohol

Types of Clouds

Wanda points out that there are three different types of clouds in the sky. The curly, wispy clouds are known as *cirrus.* The layered clouds are known as *stratus.* And the lumpy, puffy clouds are known as *cumulus.* These are the three basic types of clouds. There are, however, several variations on these basic types. For instance, any cloud with the word "nimbus" in it means it is going to produce rain!

The following illustrations show the three basic types of clouds and the *cumulonimbus* clouds. Below the illustrations, descriptions are given, along with explanations of how the clouds were named.

Cirrus clouds are high, thin, white clouds that are made of tiny ice pieces. Cirrus is a Latin word meaning curl.

Stratus clouds are low, flat gray clouds which are layered. When stratus clouds lie close to the ground, they are called fog. Stratus is a Latin word which means layer.

Cumulus clouds are white, puffy clouds which form in warm air on sunny days. They can quickly develop into thunder clouds or cumulonimbus clouds. Cumulus is Latin for heap.

Cumulonimbus clouds or thunderheads are huge, puffy, dark clouds, which are a type of cumulus cloud. Nimbus is Latin for rain.

Clouds in the Atmosphere

This table illustrates several types of clouds and the heights they can reach in the atmosphere.

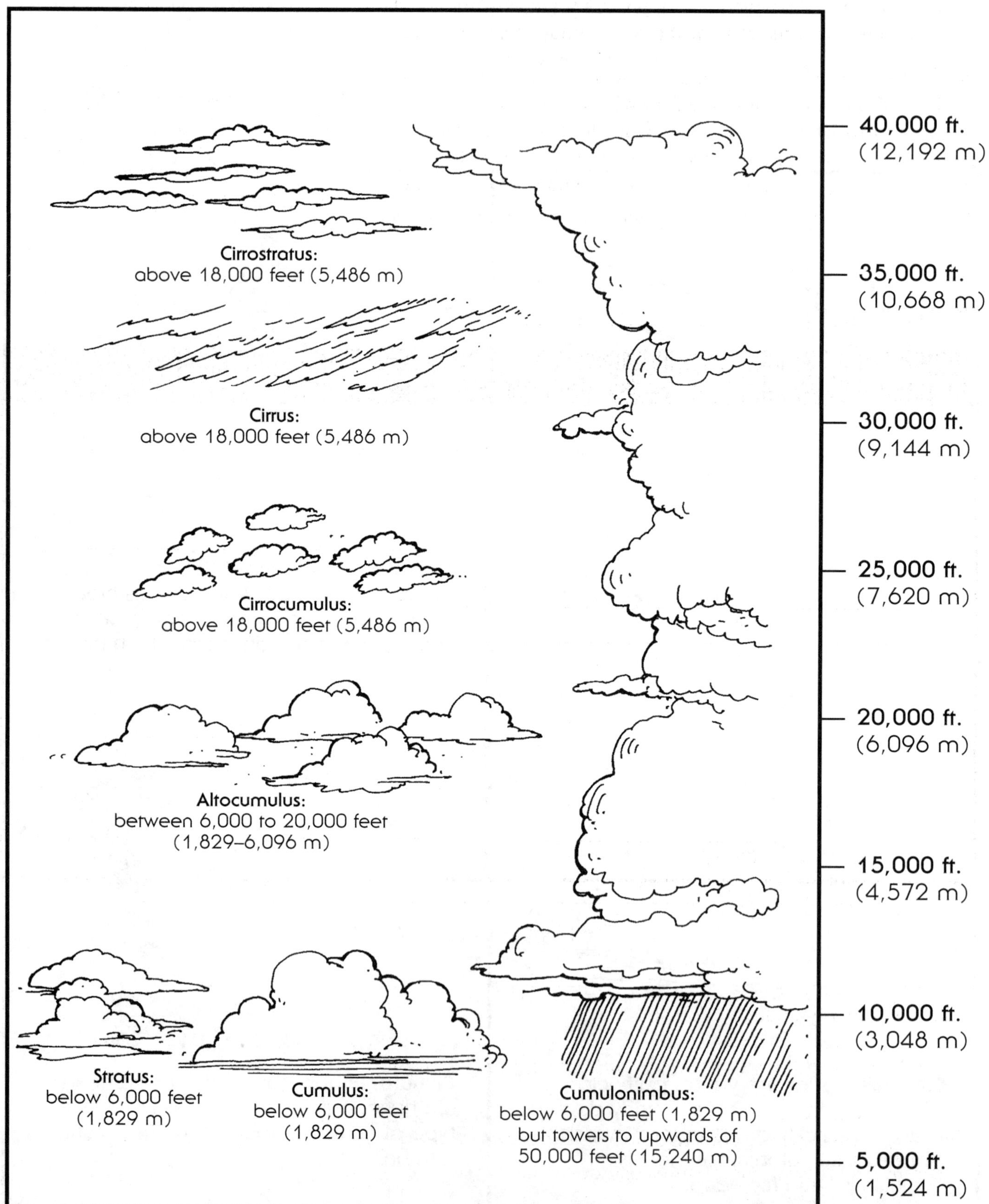

Recording Cloud Formations

Have you ever found shapes within the clouds? The act of looking for cloud shapes is called *nephelococcygia.* For this activity you will be completing a chart which chronicles the types of clouds you see in the sky and some of the shapes they seem to form.

Instructions: Look at the clouds in the sky today and see if you can identify them as one of the following: cirrus, stratus, cumulus, cirrostratus, cirrocumulus, altostratus, altocumulus, stratocumulus, or cumulonimbus. Draw the clouds and label them. In the right column, list the shapes you see in the clouds. Repeat this activity on another day when there are different types of clouds in the sky.

Cloud Chart

TODAY'S DATE ______________________

TYPES OF CLOUDS I SEE TODAY	SHAPES I SEE IN THE CLOUDS
______________________ TYPE	______________________ ______________________ ______________________ ______________________
______________________ TYPE	______________________ ______________________ ______________________ ______________________

Internet Extender

Cloud Boutique
http://vortex.plymouth.edu/clouds.html
Summary: This Web sight has excellent pictures and descriptions which help students identify clouds. It also shows *lenticular clouds* which look like flying saucers.

Clouds from Space
http://www.solarviews.com/eng/cloud1.htm#views
Summary: Outstanding pictures and descriptions of clouds as seen from space.

Weather Fronts

Ms. Frizzle and her class are now floating in the middle of a storm to learn about how rain forms. Amanda Jane and Arnold demonstrate their rainmaking project, and Ms. Frizzle explains how thousands of tiny droplets come together to form one raindrop. But what actually causes these storms in the first place are large masses of air, known as *fronts,* which collide with one another.

The layer of air which covers the earth includes a variety of temperatures. One large mass of air (a front) might be quite warm, while another front might be quite cold. When these fronts collide, our weather changes dramatically. Warm fronts tend to rise above colder fronts. The rising of warm, moist air then forms clouds and storms.

You can perform an activity which illustrates what happens when warm fronts and cold fronts collide. You will use cooking oil to simulate a warm front and colored water to simulate a cold front.

Materials:

- clear glass cooking dish
- water
- blue food dye
- cardboard
- cooking oil
- scissors

Procedure:

1. Cut the cardboard so that it forms a tight barrier between the right and left sides of the cooking dish.
2. On the right side of the barrier, pour cooking oil into the dish so that it almost fills the right side.
3. On the left side of the barrier, pour water into the dish so that it almost fills the left side. Add a few drops of blue food dye to the water.
4. When the liquids appear calm, quickly lift the barrier and watch what happens.

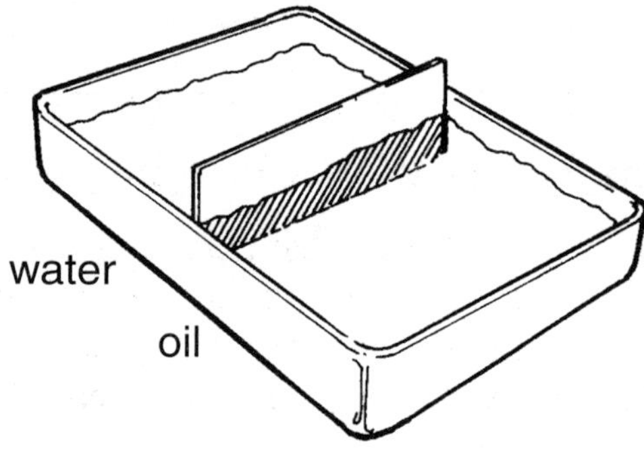

Results:

The cooking oil will rise above the colored water. Since there are no winds blowing the oil and water around, the oil will ultimately settle on the top. On earth, the warm fronts and cold fronts are in constant motion due to the winds, particularly those winds in the upper atmosphere.

Closure:

How does this activity remind you of the process of cloud formation?

____

____

____

Answer *(Note: Unless students are to self-check their responses, fold the following answers under before reproducing the page.)*

Warm, moist air must rise into the atmosphere to be subjected to lower pressure and form clouds.

Weather Fronts *(cont.)*

Cold Fronts Which Run into Warm Fronts:

Cold fronts are dense masses of cold air. When a cold front advances into a warm front, the cold front will push the lighter warm air up and out of its way. As the air of the warm front rises, it often forms cumuli or cumulonimbi. These clouds are responsible for thunderstorms. This is why thunderstorms can often be seen along the leading edge of a cold front. Cold fronts typically move in a southeasterly direction across the United States.

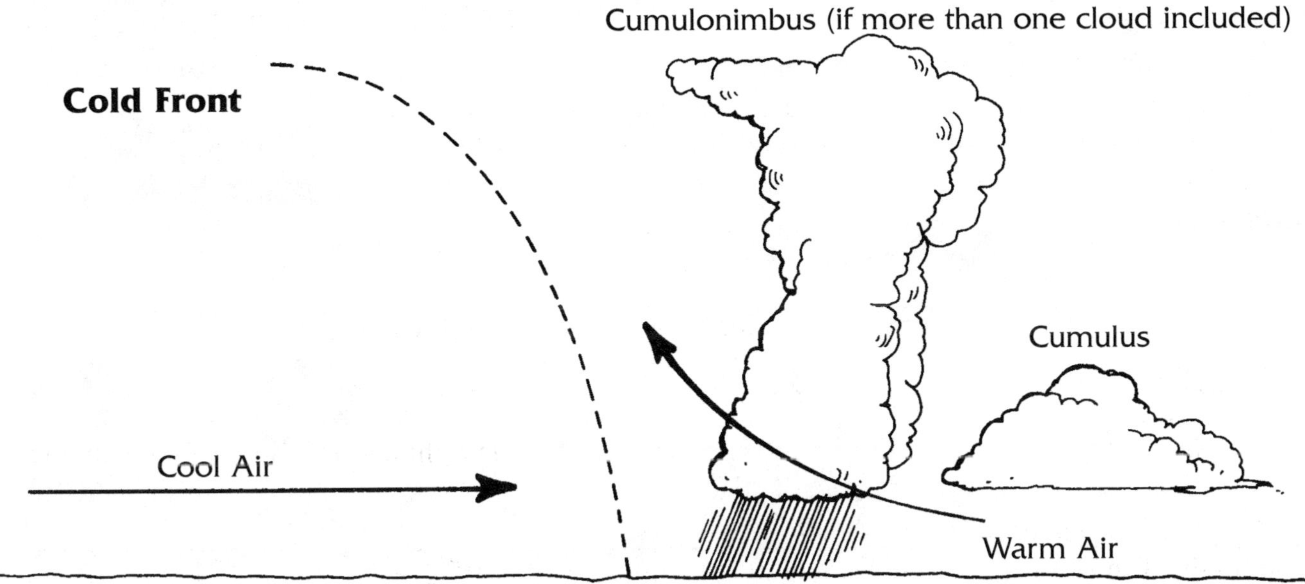

Warm Fronts Which Run into Cold Fronts:

Warm fronts are less dense than cold fronts. When a warm front runs into a cold front, the warm air is forced to rise above the cold air. This collision causes slowly rising clouds, such as cirri, altostrati, and strati. Generally, along the trailing edge of the warm front, nimbostrati are formed, which bring a drizzle or slow, steady rain to the area. Warm fronts typically move in a northeasterly direction across the United States.

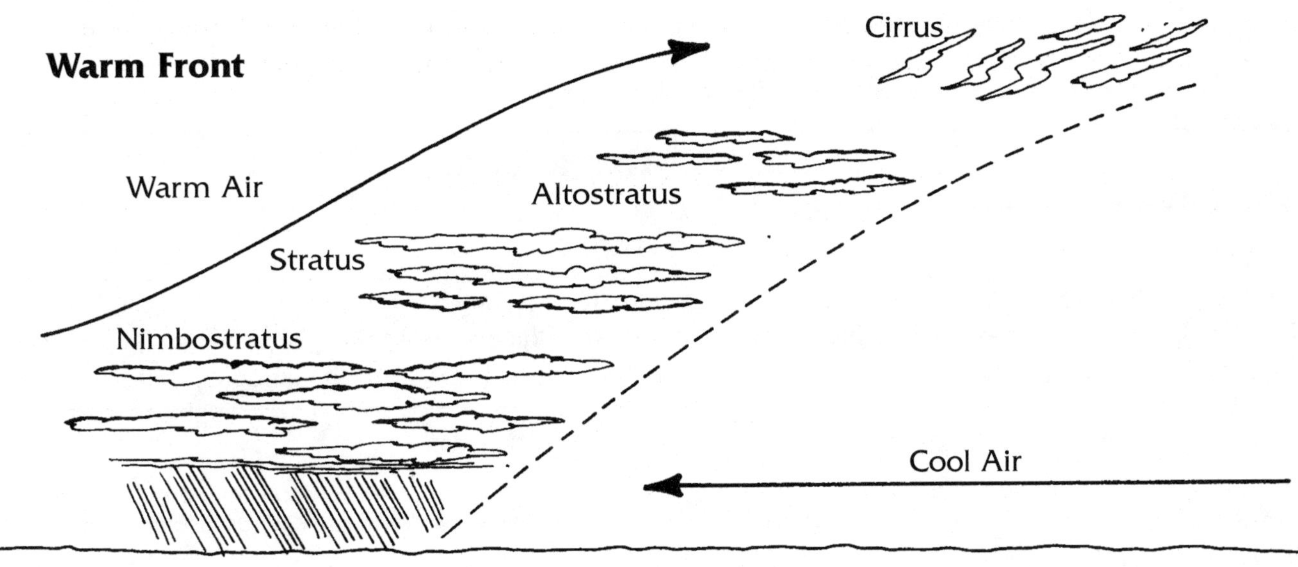

Weather Symbols

The weather service uses many symbols to plot the existing weather in the United States on a map. The following are some of these symbols which you will be using to plot (on page 39) the changes in the weather.

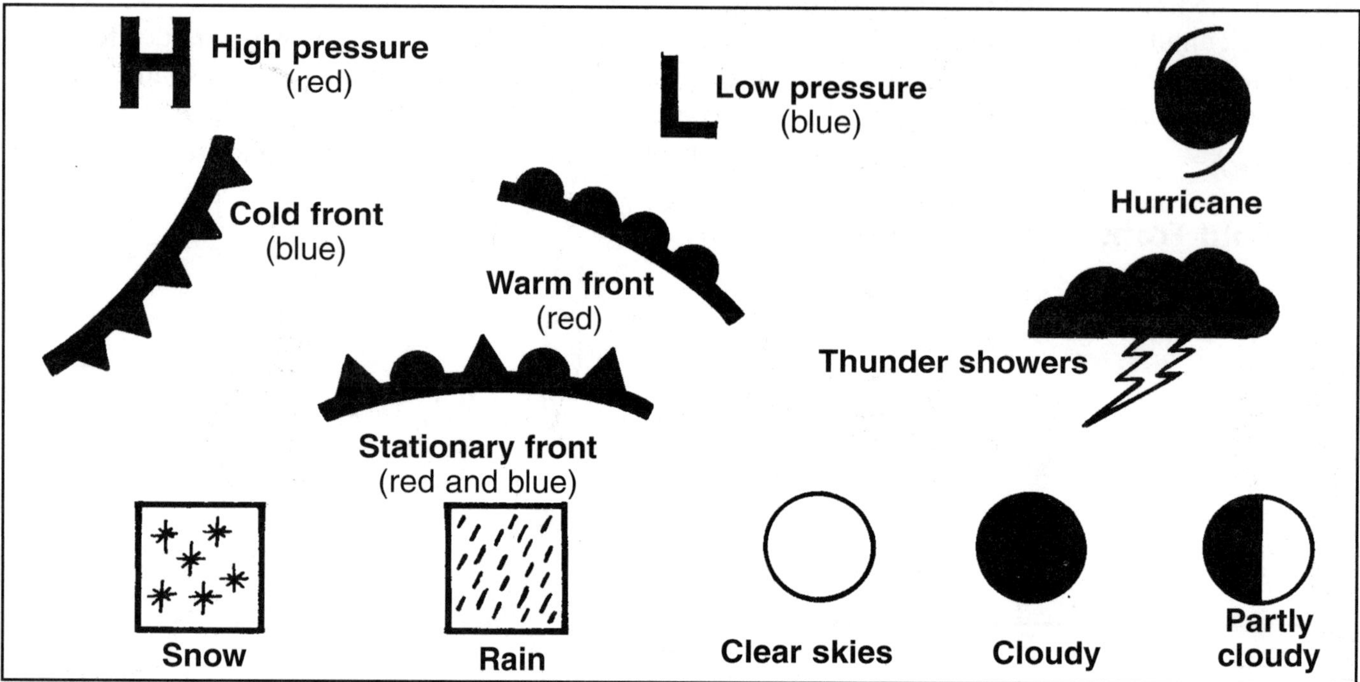

Materials:

- five copies of the map on page 39
- the weather report from a newspaper

Procedure:

Each day, for five days, use a daily weather report for the United States, along with the weather symbols, to plot the path of the weather. Lay the symbols in the appropriate areas on a map and lightly trace around them with a pencil or simply copy the symbols freehand. Use a new map each day. You will be witnessing the progress of the weather over a period of five days.

Closure:

What patterns do you notice developing as the week goes by?

Internet Extender

Yahoo Weather Forecast

http://weather.yahoo.com/

Activity Summary: Type the name of your city and state in the box and then click on search. Look at the weather information, including the forecast. Place a bookmark for this Web page and revisit it daily to see how the forecast compares with the actual observations students make.

(fold under)

Closure Answer: Students may notice cold fronts move southeasterly while warm fronts move northeasterly. Other patterns will also be noticeable, depending on the season.

Weather Symbols *(cont.)*

Outline Map of the United States

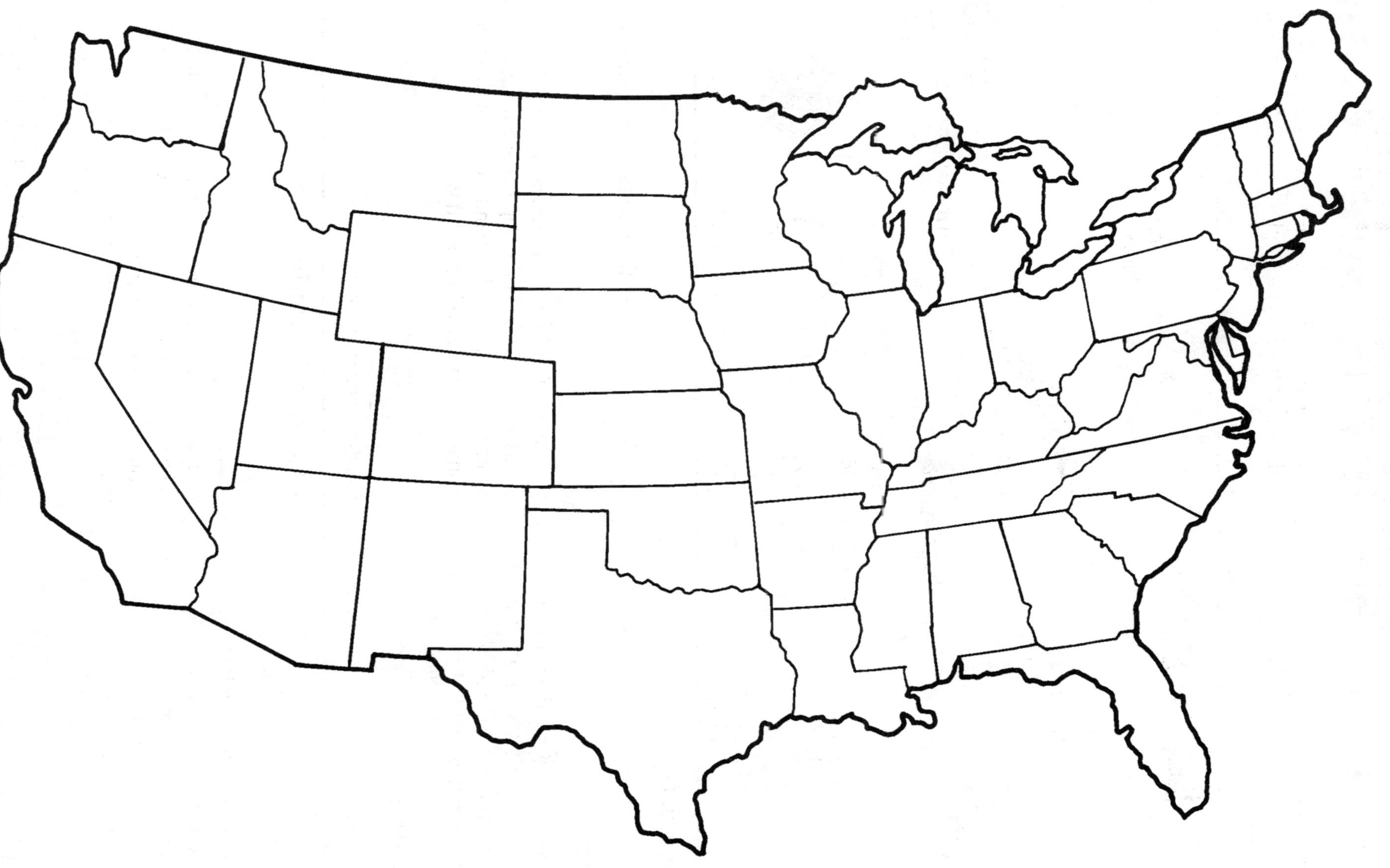

How Hurricanes Are Formed

The class soon finds themselves over a tropical ocean. They have arrived at a breeding ground for hurricanes. Hurricanes are storms which are not formed in the same way typical storms are created. Cold fronts and warm fronts do not collide to create hurricanes. Instead, hurricanes and their cousins, typhoons and cyclones, are formed in the following way:

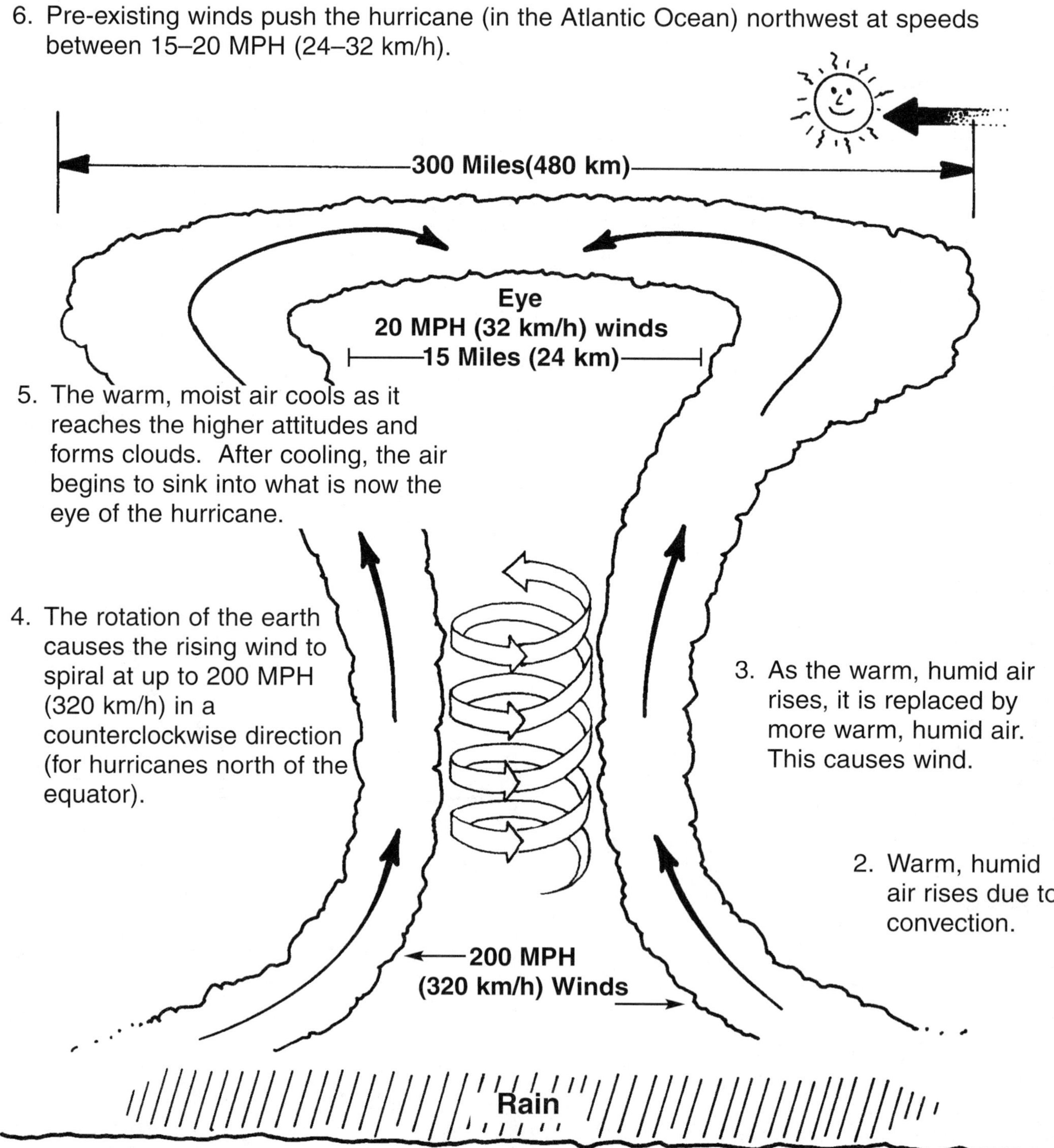

1. The ocean water must be at least 200 feet (61 m) deep and 80°F (27°C).

Name That Hurricane

They are called typhoons in the Pacific, hurricanes in the Atlantic, and meteorologists refer to them as tropical cyclones. But no matter what you call them, these ferocious storms pack a punch of winds up to 220 miles per hour (360 km/h) and can raise a wall of water 25 feet (8 m) high.

Did you know that the idea for naming hurricanes began with a book written in 1941? George R. Stewart's novel *Storm* told of a meteorologist who assigned women's names to hurricanes. It was around 1950 that forecasters adopted Stewart's idea of naming hurricanes in order to tell them apart. Today, meteorologists use a list of men and women's names for hurricanes on a rotating six-year cycle. The first hurricane of the season (which is from June to November in the North Atlantic) begins with the letter A, the second with the letter B, and so on. The names of hurricanes which cause tremendous damage when they strike land are removed from the six-year cycle.

1997	1998	1999	2000	2001	2002
Ana	Andrew	Arlene	Alberto	Allison	Arthur
Bill	Bonnie	Bret	Beryl	Barry	Bertha
Claudette	Charley	Cindy	Chris	Chantal	Cristobal
Danny	Danielle	Dennis	Debby	Dean	Dolly
Erika	Earl	Emily	Ernesto	Erin	Edoard
Fabian	Frances	Floyd	Florence	Felix	Fay
Grace	Georges	Gert	Gordon	Gabrielle	Gustav
Henri	Hermine	Harvey	Helene	Humberto	Hanna
Isabel	Ivan	Irene	Isaac	Iris	Isidore
Juan	Jeanne	Jose	Joyce	Jerry	Josephine
Kata	Karl	Katrina	Keith	Karen	Kyle
Larry	Lisa	Lenny	Leslie	Luis	Lili
Mindy	Mitch	Marie	Michael	Marilyn	Marco
Nicholas	Nicole	Nate	Nadine	Noel	Nana
Odette	Otto	Ophelia	Oscar	Opal	Omar
Peter	Paula	Philippe	Patty	Pablo	Paloma
Rose	Richard	Rita	Rafael	Roxanne	Rene
Sam	Shary	Stan	Sandy	Sebastien	Sally
Teresa	Tomas	Tammy	Tony	Tanya	Teddy
Victor	Virginie	Vince	Valerie	Van	Vicky
Wanda	Walter	Wilma	William	Wendy	Wilfred

Coriolis Effect

Due to wind patterns and the rotation of the earth, hurricanes north of the equator travel in a northwesterly direction. South of the equator, hurricanes (tropical cyclones) travel in a southeasterly direction.

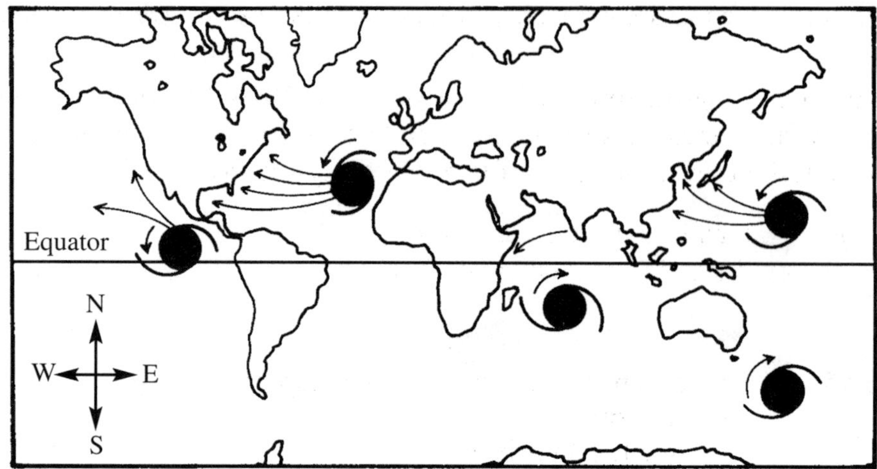

The earth's rotation also has an effect, known as the *Coriolis effect,* on the direction in which a hurricane spins. North of the equator, hurricanes spin in a counterclockwise direction. South of the equator, hurricanes (tropical cyclones) spin in a clockwise direction. The Coriolis effect will influence large weather systems and guided missile trajectories. It is, however, **not** strong enough to significantly influence the direction water spins as it goes down the drain.

You can simulate the Coriolis effect yourself by performing the following demonstration.

Materials:

- record player
- piece of cardboard
- felt-tipped marking pen

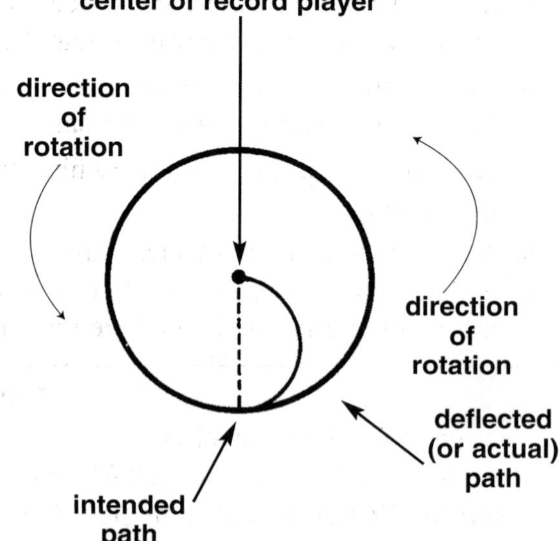

Procedure:

1. Punch a hole in the center of the cardboard so that it will fit onto the record player.
2. Turn the record player on its fastest speed. (Do not let the needle down.)
3. Using the marking pen, quickly draw a straight line from the center of the cardboard out to the edge while the player is still on.
4. Turn off the record player and look at your straight line.

Results:

The line you drew is probably not as straight as you thought it might be. The spinning cardboard, much like the spinning earth, displaces objects which attempt to travel in a straight line across its surface.

Tracking a Hurricane

Class Project

Poor Arnold is separated from the class and is in the midst of the hurricane. Meanwhile, safely back on the bus, the class is watching the progress of the hurricane as it approaches land. A hurricane, they discover, is much like a top. A top spins quickly around its axis of rotation but progresses slowly (if at all) across a table.

Wanda notes that hurricanes progress towards land at speeds of 10 to 20 miles per hour (16 to 32 kilometers per hour). In fact, hurricanes progress much faster over water than they do over land. Ultimately, a hurricane will die over land since there is not any water with which to fuel its progress.

Teacher Instructions: You and your class can track the path of a hurricane by following the directions below.

Materials:

- a copy of page 45
- colored paper
- tape
- a transparency copy of the map on page 44
- scissors
- overhead projector

Procedure:

1. Place a transparency copy of the map on page 44 on an overhead projector to project a large image on the wall.
2. Cut out a circle of colored paper which will fit over an area of 7" x 7" (18 cm x 18 cm) squares as they appear in the projected image. This is roughly the size of a large hurricane.
3. Cut out six more hurricane circles of the same size.
4. Ask for seven volunteers and hand each one a hurricane circle and some tape.
5. Begin reading the information sheet on page 45 to the class. Ask one of the volunteers to tape his/her hurricane to the wall so that the eye of the storm is located at the coordinates you have given.
6. Ask another volunteer to predict, based on what he/she has learned so far about the path of a hurricane, where the next hurricane should be placed.
7. After this volunteer has placed his/her circle in the predicted place, read the next location of the hurricane.
8. As the hurricane approaches land, ask your students when a hurricane warning should be given and for what cities. As a rule of thumb, hurricane warnings should be given at least 12 hours before the hurricane is predicted to strike an area.

Internet Extender

Tropical Cyclone Tracker
http://ww2010.atmos.uiuc.edu/(Gh)/guides/mtr/hurr/hurtrack/index.html
Activity Summary: Use the pull-down screen at this Web site to request a year between 1950 and 1998. Then pull down the screen to designate the hurricane you want to track (e.g., 1992, Andrew). The plug-in Shockwave is required for viewing the data.

Hurricanes Online
http://ww2010.atmos.uiuc.edu/(Gh)/guides/mtr/hurr/home.rxml
Activity Summary: Assign groups of students to research the various hurricane topics (e.g., Definition and Growth).

Tracking a Hurricane *(cont.)*

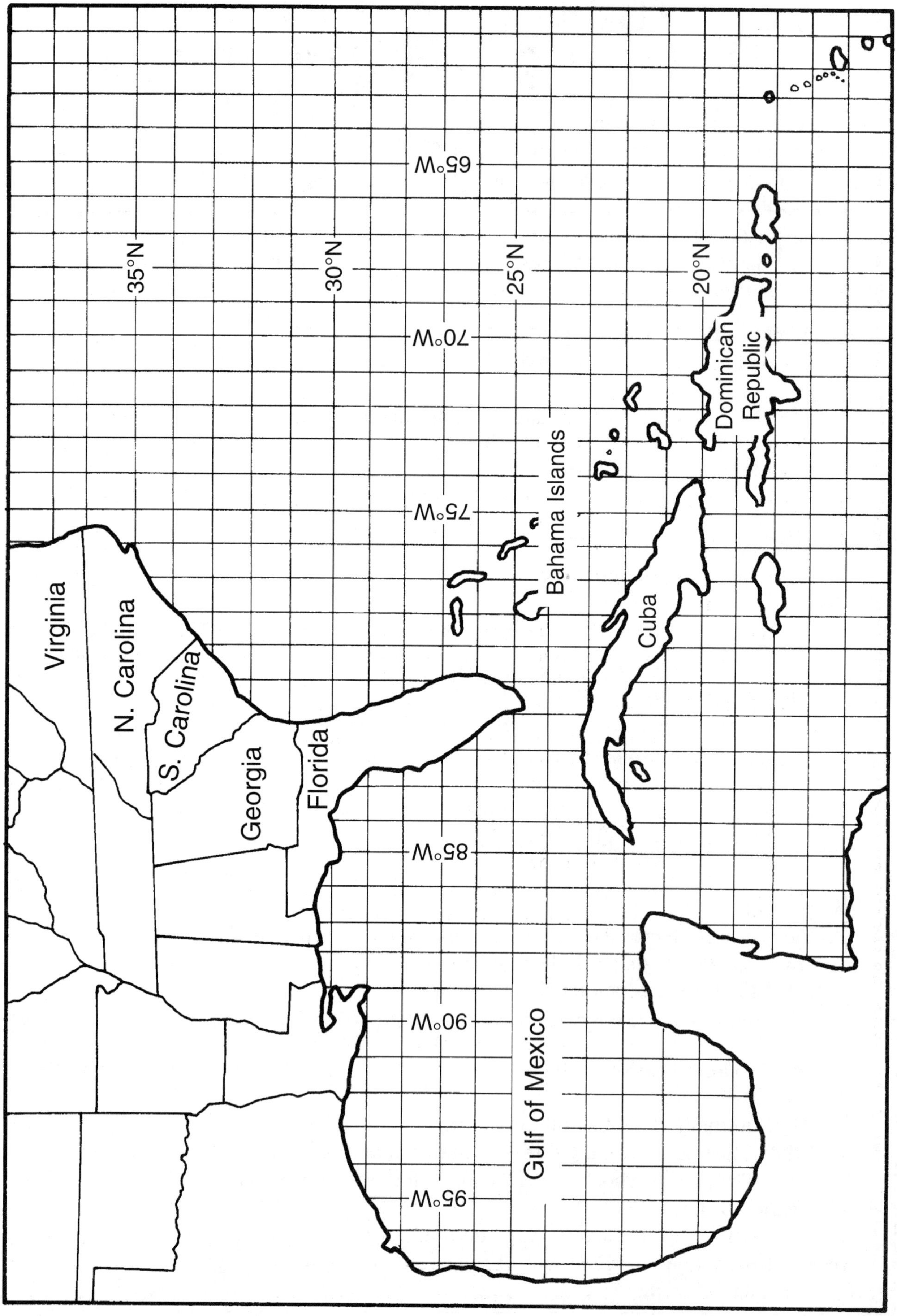

35°N
30°N
25°N
20°N
65°W
70°W
75°W
85°W
90°W
95°W
Virginia
N. Carolina
S. Carolina
Georgia
Florida
Bahama Islands
Cuba
Dominican Republic
Gulf of Mexico

Tracking a Hurricane *(cont.)*

Data Sheet

6:00 A.M., September 3:

Hurricane Arnold has just begun to form. Weather satellite images have placed its eye at 20°N and 64°W. A hurricane warning should be distributed to both the Bahamas and the Dominican Republic.

6:00 P.M., September 3:

Hurricane Arnold is picking up speed and power. It is headed straight for the Bahamas. The hurricane warning for the Bahamas was correct. Its eye can now be found at 22°N and 70°W.

6:00 A.M., September 4:

Hurricane Arnold is advancing rapidly northward. Speeds near the eye have been reported to be close to 150 MPH (241 km/h). The eye has been spotted at 24°N and 75°W. It looks like the majority of the hurricane will miss Cuba, but a hurricane warning for Miami, Florida, is in order.

6:00 P.M., September 4:

The eye of the hurricane sits just off the coast of Florida at 27°N and 79°W. Mass destruction is occurring. Forecasters are not sure of the direction the hurricane will take from here as hurricanes become unstable when they approach land. To be safe, a hurricane warning for central Florida, southern Georgia, and southern South Carolina should be given immediately.

6:00 A.M., September 5:

Hurricane Arnold has taken an abrupt turn northward and has avoided central Florida. It is still over the ocean, and its eye is centered near the border between Florida and Georgia at 31°N and 81°W. Forecasters are fairly certain the hurricane will continue in its northerly path as it is beginning to die down.

6:00 P.M., September 5:

North Carolina is feeling Hurricane Arnold just off of its coast at 34°N and 77°W. It looks at though the forecasters were correct; the hurricane is continuing up the coast, but it is dying rapidly. Winds still ravage the coastline, but the land is taking its toll on the hurricane.

6:00 A.M., September 6:

Hurricane Arnold is dying down off the coast of Virginia this morning. Rains are striking the coastline, but the fierce winds witnessed in the Bahamas and Florida are now gone. The hurricane eye can now be seen at 37°N and 75°W, but forecasters have downgraded its rating from hurricane to tropical depression. The worst is over.

What Did You Learn?

Now that you have completed reading *The Magic School Bus® Inside a Hurricane* and have taken part in the unit activities, it is time to assess your understanding of the weather.

Assessment 1: The Weather

Using many of the skills you acquired throughout this science/literature unit, you will now keep track of the weather in your area for several days.

Procedure:

1. Make a copy of the charts on page 47.
2. Observe the weather for the days indicated on the charts.
3. Record the date and time you make your measurements in the first two columns.
4. In Chart A, use the instruments you built or acquired during this unit to fill in the following areas:

    Temperature (Use a thermometer.)

    Barometer (Remember to note whether the barometer is rising, falling, or stable. Use a newspaper or television weather report to find the barometric pressure.)

    Wind Direction (Use a homemade weather vane and a compass.)

    Percent Humidity (Use your homemade sling psychrometer and the table on page 29.)

5. In Chart B, describe the current weather conditions and predict what the weather may look like tomorrow. There is a box on the extreme right-hand side of the chart to mark "yes" if your prediction was correct or "no" if your prediction was not correct. Use the terms from the word box below to describe the current weather conditions and to predict the future conditions or use any other words you feel are appropriate to describe the weather.

Word Box

raining	hailing	partly cloudy	smoggy
snowing	cloudy	sunny	foggy

Assessment 2: Tracking a Hurricane

Using the Tracking a Hurricane activity on pages 43–45 as a guide, you will rewrite the story of Hurricane Arnold.

Procedure:

1. Follow the instructions on page 43 for making the map and hurricane symbols.
2. Place the seven hurricane symbols on the map in a pattern which would follow the path of a hurricane. (Remember that hurricanes north of the equator travel in a northwesterly direction. Be sure your path is somewhat northwesterly.)
3. Look at the map and write down the seven different coordinates of the hurricane.
4. Name the hurricane and write a story about its progress and its areas of destruction.

What Did You Learn? *(cont.)*

Assessment 1: The Weather *(cont.)*

Instructions: Use these two charts to record your data about the weather for several days. In the first chart, record raw data about the weather. In the second chart, record how the weather looks today and your predictions for tomorrow. At the end of the second chart, write "Y" or "N" to indicate the outcome of your prediction. You will not need to predict the weather on Saturday; only confirm, on the line above, if your prediction was correct or not.

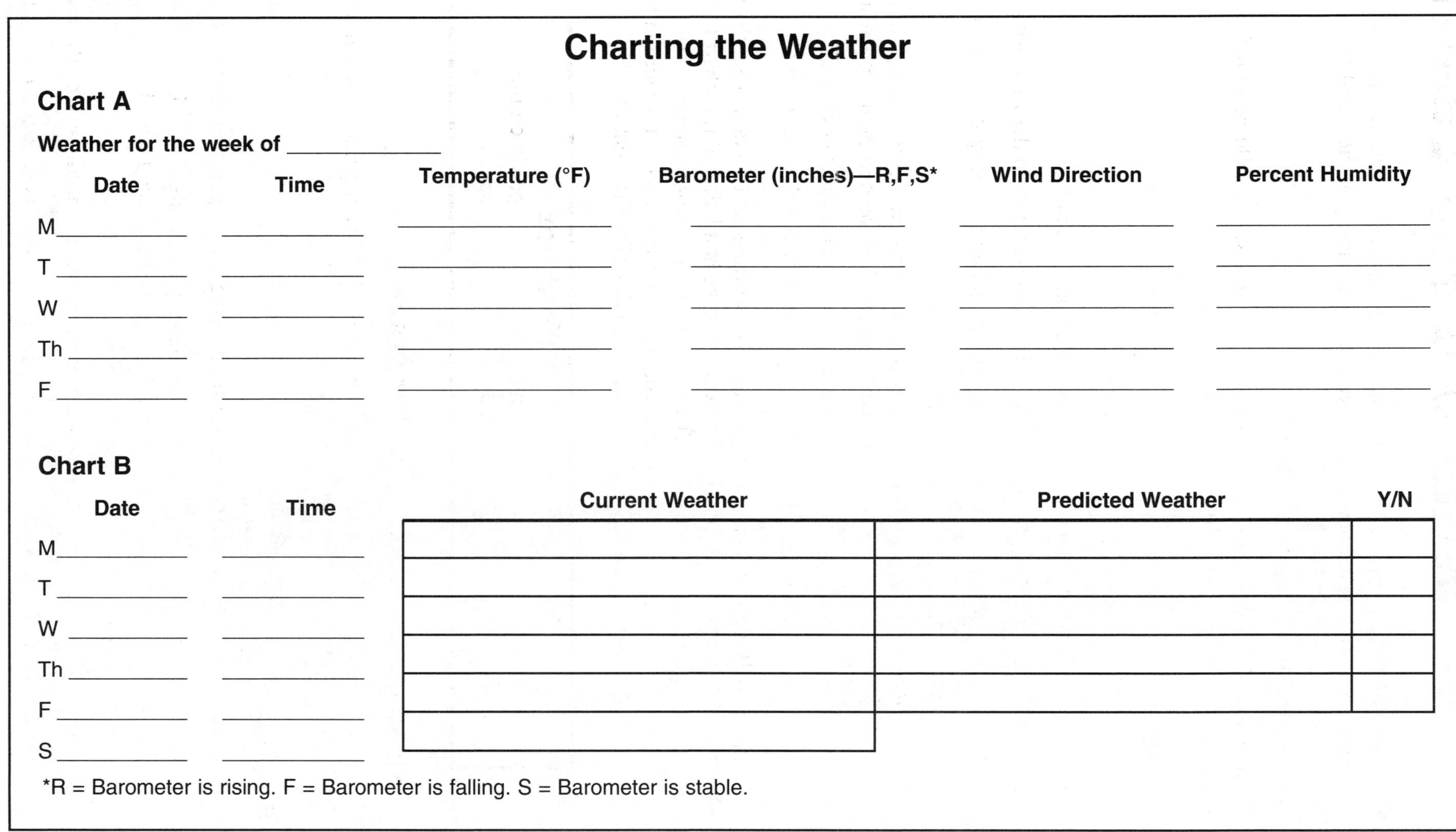

Charting the Weather

Chart A

Weather for the week of ____________

Date	Time	Temperature (°F)	Barometer (inches)—R,F,S*	Wind Direction	Percent Humidity
M					
T					
W					
Th					
F					

Chart B

Date	Time	Current Weather	Predicted Weather	Y/N
M				
T				
W				
Th				
F				
S				

*R = Barometer is rising. F = Barometer is falling. S = Barometer is stable.

Related Books and Materials

Related Books

Cole, Joanna. *The Magic School Bus® Inside a Hurricane.* Scholastic, Inc., 1995.
This delightful book takes the reader on an adventure with Ms. Frizzle's class to the inside of a hurricane. It is a fiction book which includes scientifically accurate descriptions of the behavior of the weather.

Cosgrove, Brian. *Eyewitness Books: Weather.* Alfred A. Knopf, 1991.
This colorfully illustrated book explores the principles of weather.

Couchman, J.K. et al. *Mini Climates.* Holt, Rinehart, and Winston of Canada, 1971.
A fabulous book, it details hands-on activities dealing with weather and climate.

Day, John and Vincent Schaefer. *Clouds and Weather: A Peterson First Guide.* Houghton Mifflin, 1991.
This is a good introductory guide to the weather.

Forrester, Frank. *One Thousand One Questions Answered About the Weather.* Dover, 1982.
A reference guide, it answers many questions about the weather.

Liem, Tik L. *Invitations to Science Inquiry.* Science Inquiry Enterprises, 1987.
Dr. Liem's book is a comprehensive text of activities and demonstrations for any classroom teacher interested in science. *Invitations to Science Inquiry* is perhaps the best book available for activity based science.
(This book can be ordered through the NSTA catalog or directly from Science Inquiry Enterprises at 14358 Village View Lane, Chino Hills, CA 91709.)

Mogil, Michael and Barbara Levine. *The Amateur Meteorologist.* Franklin Watts, 1993.
This excellent book includes explorations and investigations into the science of meteorology.

Williams, Jack. *The Weather Book.* USA Today, 1992.
A "must have" book for any classroom, it has Colorful illustrations and simple explanations of the fundamentals of weather.

Zimmerman, Barry and David. *Why Nothing Can Travel Faster Than Light . . .* Contemporary Books, 1993.
A wonderful book of essays, it deals with many questions asked about topics in science. One essay is devoted to the explanation of hurricanes.

Related Materials

Acorn Naturalists, 17300 East 17th Street, #J-236, Tustin, CA 92680. (800) 422-8886
This company supplies activity and reference books and materials in all science areas.

American Meteorological Society (AMS), 1701 K Street NW, Suite 300, Washington, DC 20006.
The American Meteorological Society distributes educational literature related to weather and climate. Write to them for more details.

Association of American Weather Observers (AAWO), P.O. Box 455, Belvedere, IL 61008.
The AAWO maintains a speakers' bureau and a library on weather related topics. They also publish a monthly periodical, *The American Weather Observer,* which covers topics in weather and climate.

The Magic School Bus® Explores the Ocean. Software. Microsoft. (800) 376-5125
Microsoft has adapted several books in *The Magic School Bus®* series into computer adventures. These are on CD-ROM for Windows and Macintosh.

National Climatic Data Center, Federal Building, Asheville, NC 28801. The National Climate Data Center distributes educational literature related to weather and climate. Write to them for more details.

National Science Teacher Association (NSTA), 1840 Wilson Blvd., Arlington, VA 22201-3000. (800) 722-NSTA
Request a catalog of their classroom science materials.

National Weather Association (NWA), 501 Capitol Ct. NE Ste. 100, Washington, DC 20002.
The NWA publishes the following materials on weather: *A Comprehensive Glossary of Weather Terms for Storm Spotters,* cloud charts, and a quarterly journal entitled the *National Weather Digest.*

National Weather Service Public Affairs Office, 1325 East-West Highway, Silver Spring, MD 20910.
The National Weather Service Public Affairs Office distributes educational literature related to weather and climate. Write to them for more details.

PBS Home Video, Pacific Arts, Nesmith Enterprises Inc., 11858 LaGrange Ave., Los Angeles, CA 90025.
PBS provides the Reading Rainbow video series, which includes some of *The Magic School Bus®* series.

